The

THE DESCENT OF WOMAN

Elaine Morgan

SOUVENIR PRESS

Contents

Acknowledgements

The quotations from Sir Alister Hardy and Desmond Morris, appearing on pages 276 and 277, are taken from the script of the documentary film *The Water Babies* by kind permission of Golden Dolphin Productions Ltd.

One

The Man-made Myth

According to the Book of Genesis, God first created man. Woman was not only an afterthought, but an amenity. For close on two thousand years this holy scripture was believed to justify her subordination and explain her inferiority; for even as a copy she was not a very good copy. There were differences. She was not one of His best efforts.

There is a line in an old folk song that runs: 'I called my donkey a horse gone wonky.' Throughout most of the literature dealing with the differences between the sexes there runs a subtle underlying assumption that woman is a man gone wonky; that woman is a distorted version of the original blueprint; that they are the norm, and we are the deviation.

It might have been expected that when Darwin came along and wrote an entirely different account of *The Descent of Man*, this assumption would have been eradicated; for Darwin didn't believe she was an afterthought: he believed her origin was at least contemporaneous with man's. It should have led to some kind of breakthrough in the relationship between the sexes. But it didn't.

Almost at once men set about the congenial and fascinating task of working out an entirely new set of reasons why woman was manifestly inferior and irreversibly subordinate, and they have been happily engaged on this ever since. Instead of theology they use biology, and ethology, and primatology, but they use them to reach the same conclusions.

They are now prepared to debate the most complex

problems of economic reform not in terms of the will of God, but in terms of the sexual behaviour patterns of the cichlid fish; so that if a woman claims equal pay or the right to promotion there is usually an authoritative male thinker around to deliver a brief homily on hormones, and point out that what she secretly intends by this, and what will inevitably result, is the 'psychological castration' of the men in her life.

Now, that may look to us like a stock piece of emotional blackmail—like the woman who whimpers that if Sonny doesn't do as she wants him to do, then Mother's going to have one of her nasty turns. It is not really surprising that most women who are concerned to win themselves a new and better status in society tend to sheer away from the whole subject of biology and origins, and hope that we can ignore all that and concentrate on ensuring that in the future things will be different.

I believe this is a mistake. The legend of the jungle heritage and the evolution of man as a hunting carnivore has taken root in man's mind as firmly as Genesis ever did. He may even genuinely believe that equal pay will do something terrible to his gonads. He has built a beautiful theoretical construction, with himself on the top of it, buttressed with a formidable array of scientifically authenticated facts. We cannot dispute the facts. We should not attempt to ignore the facts. What I think we can do is to suggest that the currently accepted interpretation of the facts is not the only possible one.

I have considerable admiration for scientists in general, and evolutionists and ethologists in particular, and though I think they have sometimes gone astray, it has not been purely through prejudice. Partly it is due to sheer semantic accident—the fact that 'man' is an ambiguous term. It means the species; it also means the male of the species. If you begin to write a book about man or conceive a theory about man you cannot avoid using this word. You

cannot avoid using a pronoun as a substitute for the word,
and you will use the pronoun 'he' as a simple matter of
linguistic convenience. But before you are halfway through
the first chapter a mental image of this evolving creature
begins to form in your mind. It will be a male image,
and he will be the hero of the story: everything and everyone else in the story will relate to him.

All this may sound like a mere linguistic quibble or a
piece of feminist petulance. If you stay with me, I hope to
convince you it's neither. I believe the deeply rooted
semantic confusion between 'man' as a male and 'man' as
a species has been fed back into and vitiated a great deal
of the speculation that goes on about the origins, development, and nature of the human race.

A very high proportion of the thinking on these topics
is androcentric (male-centred) in the same way as pre-
Copernican thinking was geocentric. It's just as hard for
man to break the habit of thinking of himself as central
to the species as it was to break the habit of thinking of himself as central to the universe. He sees himself quite unconsciously as the main line of evolution, with a female satellite
revolving around him as the moon revolves around the
earth. This not only causes him to overlook valuable clues
to our ancestry, but sometimes leads him into making
statements that are arrant and demonstrable nonsense.

The longer I went on reading his own books about
himself, the more I longed to find a volume that would
begin: 'When the first ancestor of the human race descended from the trees, she had not yet developed the
mighty brain that was to distinguish her so sharply from
all other species. . . .'

Of course, she was no more the first ancestor than he
was—but she was no *less* the first ancestor, either. She was
there all along, contributing half the genes to each suc-
ceeding generation. Most of the books forget about her
for most of the time. They drag her onstage rather sud-

denly for the obligatory chapter on Sex and Reproduction, and then say: 'All right, love, you can go now,' while they get on with the real meaty stuff about the Mighty Hunter with his lovely new weapons and his lovely new straight legs racing across the Pleistocene plains. Any modifications in her morphology are taken to be imitations of the Hunter's evolution, or else designed solely for his delectation.

Evolutionary thinking has been making great strides lately. Archaeologists, ethologists, paleontologists, geologists, chemists, biologists and physicists, are closing in from all points of the compass on the central area of mystery that remains. For despite the frequent triumph dances of researchers coming up with another jawbone or another statistic, some part of the miracle is still unaccounted for. Most of their books include some such phrase as: '... the early stages of man's evolutionary progress remain a total mystery.' 'Man is an accident, the culmination of a series of highly improbable coincidences....' 'Man is a product of circumstances special to the point of disbelief.' They feel there is still something missing, and they don't know what.

The trouble with specialists is that they tend to think in grooves. From time to time something happens to shake them out of that groove. Robert Ardrey tells how such enlightenment came to Dr. Kenneth Oakley when the first Australopithecus remains had been unearthed in Africa: 'The answer flashed without warning in his own large-domed head, "Of course we believed that the big brain came first! We assumed that the first man was an Englishman!"' Neither he, nor Ardrey in relating the incident, noticed that he was still making an equally unconscious, equally unwarrantable assumption. One of these days an evolutionist is going to strike a palm against his large-domed head and cry: 'Of course! We assumed the first human being was a man!'

* * *

First, let's have a swift recap of the story as currently related, for despite all the new evidence recently brought to light, the generally accepted picture of human evolution has changed very little.

Smack in the centre of it remains the Tarzanlike figure of the prehominid male who came down from the trees, saw a grassland teeming with game, picked up a weapon, and became a Mighty Hunter.

Almost everything about us is held to have derived from this. If we walk erect, it was because the Mighty Hunter had to stand tall to scan the distance for his prey. If we lived in caves it was because hunters need a base to come home to. If we learned to speak it was because hunters need to plan the next safari and boast about the last. Desmond Morris, pondering on the shape of a woman's breasts, instantly deduced that they evolved because her mate became a mighty hunter, and defends this preposterous proposition with the greatest ingenuity. There's something about the Tarzan figure which has them all mesmerized.

I find the whole yarn pretty incredible. It is riddled with mysteries, and inconsistencies, and unanswered questions. Even more damning than the unanswered questions are the questions that are never even asked, because, as Professor Peter Medawar has pointed out, 'scientists tend not to ask themselves questions until they can see the rudiments of an answer in their minds'. I shall devote this chapter to pointing out some of these problems before outlining a new version of the Naked Ape story which will suggest at least possible answers to every one of them, and fifteen or twenty others besides.

The first mystery is, 'What happened during the Pliocene?'

There is a wide acceptance now of the theory that the human story began in Africa. Twenty million years ago in Kenya, there existed a flourishing population of apes of generalized body structure and of a profusion of types from the size of a small gibbon up to that of a large gorilla. Dr. L. S. B. Leakey has dug up their bones by the hundred in the region of Lake Victoria, and they were clearly doing very well there at the time. It was a period known as the Miocene. The weather was mild, the rainfall was heavier than today, and the forests were flourishing. So far, so good.

Then came the Pliocene drought. Robert Ardrey writes of it: 'No mind can apprehend in terms of any possible human experience the duration of the Pliocene. Ten desiccated years were enough, a quarter of a century ago, to produce in the American Southwest that maelstrom of misery, the dust bowl. To the inhabitant of the region the ten years must have seemed endless. But the African Pliocene lasted twelve million.'

On the entire African continent no Pliocene fossil bed of pre-human remains has ever been found. During this period many promising Miocene ape species were, not surprisingly, wiped out altogether. A few were trapped in dwindling pockets of forest and when the Pliocene ended they reappeared as brachiating apes—specialized for swinging by their arms.

Something astonishing also reappeared—the Australopithecines, first discovered by Professor Raymond Dart in 1925 and since unearthed in considerable numbers.

Australopithecus emerged from his horrifying twelve-million-year ordeal much refreshed and improved. The occipital condyles of his skull show that he carried his head at a new angle, suggesting that he walked upright rather than on all fours, and the orbital region, according to Sir Wilfred le Gros Clark, has 'a remarkably human appearance'. He was clever, too. His remains have been found in the Olduvai Gorge in association with crude pebble tools that have been hailed as the earliest beginning of human culture. Robert Ardrey says: 'We entered the [Pliocene]

crucible a generalized creature bearing only human potential. We emerged a being lacking only a proper brain and a
chin. What happened to us along the way?' The sixty-
four-thousand-dollar question: 'What happened to them?
Where did they go?'

Second question: 'Why did they stand upright?' The
popular versions skim very lightly over this patch of thin
ice. Desmond Morris says simply: 'With strong pressure
on them to increase their prey-killing prowess, they became more upright—fast, better runners.' Robert Ardrey
says equally simply: 'We learned to stand erect in the
first place as a necessity of the hunting life.'

But wait a minute. We were quadrupeds. These statements imply that a quadruped suddenly discovered that
he could move faster on two legs than on four. Try to
imagine any other quadruped discovering that—a cat?
a dog? a horse?—and you'll see that it's totally non-
sensical. Other things being equal, four legs are bound to
run faster than two. The bipedal development was violently unnatural.

Stoats, gophers, rabbits, chimpanzees, will sit or stand
bipedally to gaze into the distance, but when they want
speed they have sense enough to use all the legs they've
got. The only quadrupeds I can think of that can move
faster on two legs than four are things like kangaroos—
and a small lizard called the Texas boomer, and he
doesn't keep it up for long. The secret in these cases is a
long heavy counterbalancing tail which we certainly never
had. You may say it was a natural development for a
primate because primates sit erect in trees—but *was* it
natural? Baboons and macaques have been largely terrestial for millions of years without any sign of becoming
bipedal.

George A. Bartholomew and Joseph B. Birdsell point
out: '... the extreme rarity of bipedalism among animals
suggests that it is inefficient except under very special

circumstances. Even modern man's unique vertical locomotion when compared to that of quadrupedal mammals,
is relatively ineffective.... A significant nonlocomotor advantage must have resulted.'

What was this advantage? The Tarzanists suggest that
bipedalism enabled this ape to race after game while
carrying weapons—in the first instance, presumably pebbles. But a chimp running off with a banana (or a pebble)
if he can't put it in his mouth, will carry it in one hand
and gallop along on the others, because even *three* legs
are faster than two. So what was our ancestor supposed to
be doing? Shambling along with a rock in each hand?
Throwing boulders that took two hands to lift? Tossing
cabers?

No. There must have been a pretty powerful reason
why we were constrained over a long period of time to
walk about on our hind legs *even though it was slower.*
We need to find that reason.

Third question: How did the ape come to be using
these weapons, anyway? Again Desmond Morris clears this
one lightly, at a bound: 'With strong pressure on them
to increase their prey-killing prowess ... their hands became strong efficient weapon-holders.' Compared to
Morris, Robert Ardrey is obsessed with weapons which he
calls 'mankind's most significant cultural endowment'.
Yet his explanation of how it all started is as cursory as
anyone else's: 'In the first evolutionary hour of the human
emergence we became sufficiently skilled in the use of
weapons to render redundant our natural primate daggers' (i.e. the large prehominid canine teeth).

But wait a minute—how? and why? Why did one, and
only one, species of those Miocene apes start using
weapons? A cornered baboon will fight a leopard; a
hungry baboon will kill and eat a chicken. He could
theoretically pick up a chunk of flint and forget about his
'natural primate daggers', and become a Mighty Hunter.

He doesn't do it, though. Why did we? Sarel Eimerl and Irven de Vore point out in their book *The Primates*:

'Actually, it takes quite a lot of explaining. For example, if an animal's normal mode of defence is to flee from a predator, it flees. If its normal method of defence is to fight with its teeth, it fights with its teeth. It does not suddenly adopt a totally new course of action, such as picking up a stick or a rock and throwing it. The idea would simply not occur to it, and even if it did, the animal would have no reason to suppose that it would work.'

Now primates do acquire useful tool-deploying habits. A chimpanzee will use a stick to extract insects from their nests; and a crumpled leaf to sop up water. Wolfgang Köhler's apes used sticks to draw fruit towards the bars of their cage, and so on.

But this type of learning depends on three things. There must be leisure for trial-and-error experiment. The tools must be either in unlimited supply (a forest is full of sticks and leaves) or else in *exactly the right place*. (Even Köhler's brilliant Sultan could be stumped if the fruit was in front of him and a new potential tool was behind him—he needed them both in view at the same time.) Thirdly, for the habit to stick, the same effect must result from the same action every time.

Now look at that ape. The timing is wrong—when he's faced with a bristling rival or a charging cat or even an escaping prey, he won't fool around inventing fancy methods. A chimp sometimes brandishes a stick to convey menace to an adversary, but if his enemy keeps coming, he drops the stick and fights with hands and teeth. Even if we postulate a mutant ape cool enough to think, with the adrenalin surging through his veins, 'There must be a better way than teeth,' he still has to be lucky to notice that right in the middle of the primeval grassland there happens to be a stone of convenient size, precisely between him and his enemy. And when he throws it, he has to score a bull's-eye, first time and every time. Because if he failed to hit a leopard he wouldn't be there to tell his progeny that the trick only needed polishing up a bit; and if he failed to hit a springbok he'd think: 'Ah well, that obviously doesn't work. Back to the old drawing board.'

No. If it had taken all that much luck to turn man into a killer, we'd all be still living on nut cutlets.

A lot of Tarzanists privately realize that their explanations of bipedalism and weapon-wielding won't hold water. They have invented the doctrine of 'feedback', which states that though these two theories are separately and individually nonsense, together they will just get by. It is alleged that the ape's bipedal gait, however unsteady, made him a better rock thrower (why?) and his rock throwing, however inaccurate, made him a better biped (Why?) Eimerl and de Vore again put the awkward question: Since chimps can both walk erect and manipulate simple tools, 'why was it only the hominids who benefited from the feed-back?' You may well ask.

Next question: Why did the naked ape become naked?

Desmond Morris claims that, unlike more specialized carnivores such as lions and jackals, the ex-vegetarian ape was not physically equipped to 'make lightning dashes after his prey'. He would 'experience considerable overheating during the hunt, and the loss of body hair would be of great value for the supreme moments of the chase'.

This is a perfect example of androcentric thinking. There were two sexes around at the time, and I don't believe it's ever been all that easy to part a woman from a fur coat, just to save the old man from getting into a muck-sweat during his supreme moments. What was supposed to be happening to the female during this period of de-nudation?

Dr. Morris says: 'This system would not work, of course, if the climate was too intensely hot, because of damage

to the exposed skin.' So he is obviously dating the loss of
hair later than the Pliocene 'inferno'. But the next period
was the turbulent Pleistocene, punctuated by mammoth
African 'pluvials', corresponding to the Ice Ages of the
north. A pluvial was century after century of torrential
rainfall; so we have to picture our maternal ancestor
sitting naked in the middle of the plain while the heavens
emptied, needing both hands to keep her muddy grip on
a slippery, squirming, equally naked infant. This is ludicrous. It's no advantage to the species for the mighty
hunter to return home safe and cool if he finds his son's
been dropped on his head and his wife is dead of hypo-
thermia.

This problem could have been solved by dimorphism—
the loss of hair could have gone further in one sex than
the other. So it did, of course. But unfortunately for the
Tarzanists it was the stay-at-home female who became
nakedest, and the overheated hunter who kept the hair
on his chest.

Next question: Why has our sex life become so involved and confusing?

The given answer, I need hardly say, is that it all began
when man became a hunter. He had to travel long distances after his prey and he began worrying about what
the little woman might be up to. He was also anxious
about other members of the hunting pack, because,
Desmond Morris explains, 'if the weaker males were going
to be expected to co-operate on the hunt, they had to be
given more sexual rights. The females would have to be
more shared out.'

Thus it became necessary, so the story goes, to establish
a system of 'pair bonding' to ensure that couples remained faithful for life. I quote: 'The simplest and most
direct method of doing this was to make the shared activities of the pair more complicated and more rewarding.
In other words, to make sex sexier.'

To this end, the naked apes sprouted ear lobes, fleshy nostrils, and everted lips, all allegedly designed to stimulate one another to a frenzy. Mrs. A.'s nipples became highly erogenous, she invented and patented the female orgasm, and she learned to be sexually responsive at all times, even during pregnancy, 'because with a one-male— one-female system, it would be dangerous to frustrate the male for too long a period. It might endanger the pair bond.' He might go off in a huff, or look for another woman. Or even refuse to co-operate on the hunt.

In addition, they decided to change over to face-to-face sex, instead of the male mounting from behind as previously, because this new method led to 'personalized sex'. The frontal approach means that 'the incoming sexual signals and rewards are kept tightly linked with the identity signals from the partner'. In simpler words, you know who you're doing it with.

This landed Mrs. Naked Ape in something of a quandary. Up till then, the fashionable thing to flaunt in sexual approaches had been 'a pair of fleshy, hemispherical buttocks'. Now all of a sudden they were getting her nowhere. She would come up to her mate making full-frontal identity signals like mad with her nice new earlobes and nostrils, but somehow he just didn't want to know. He missed the fleshy hemispheres, you see. The position was parlous, Dr. Morris urges. 'If the female of our species was going to successfully shift the interest of the male round to the front, evolution would have to do something to make the frontal region more stimulating.' Guess what? Right first time: she invested in a second pair of fleshy hemispheres in the thoracic region and we were once more saved by the skin of our teeth.

All this is good stirring stuff, but hard to take seriously. Wolf packs manage to co-operate without all this erotic paraphernalia. Our near relatives the gibbons remain faithful for life without 'personalized' frontal sex, without elaborate erogenous zones, without perennial female availability. Why couldn't we?

Above all, since when has increased sexiness been a guarantee of increased fidelity? If the naked ape could see all this added sexual potential in his own mate, how could he fail to see the same thing happening to all the other females around him? What effect was that supposed to have on him, especially in later life when he noticed Mrs. A.'s four hemispheres becoming a little less fleshy than they used to be?

We haven't yet begun on the unasked questions. Before ending this chapter I will mention just two out of many.

First: If female orgasm was evolved in our species for the first time to provide the woman with a 'behavioural reward' for increased sexual activity, why in the name of Darwin has the job been so badly bungled that there have been whole tribes and whole generations of women hardly aware of its existence? Even in the sex-conscious U.S.A., according to Dr. Kinsey, it rarely gets into proper working order before the age of about thirty. How could natural selection ever have operated on such a rickety, unreliable, late-developing endowment when in the harsh conditions of prehistory a woman would be lucky to survive more than twenty-nine years, anyway?

Second: Why in our species has sex become so closely linked with aggression? In most of the higher primates sexual activity is the one thing in life which is totally incompatible with hostility. A female primate can immediately deflect male wrath by presenting her backside and offering sex. Even a male monkey can calm and appease a furious aggressor by imitating the gesture. Nor is the mechanism confined to mammals. Lorenz tells of an irate lizard charging down upon a female painted with male markings to deceive him. When he got close enough to realize his mistake, the taboo was so immediate and so absolute that his aggression went out like a light, and being too late to stop himself he shot straight up into the air and turned a back somersault.

Female primates admittedly are not among the species that can count on this absolute chivalry at all times. A female monkey may be physically chastized for obstreperous behaviour; or a male may (on rare occasions) direct hostility against her when another male is copulating with her; but between the male and female engaged in it, sex is always the friendliest of interactions. There is no more hostility associated with it than with a session of mutual grooming.

How then have sex and aggression, the two irreconcilables of the animal kingdom, become in our species alone so closely interlinked that the words for sexual activity are spat out as insults and expletives? In what evolutionary terms are we to explain the Marquis de Sade, and the subterranean echoes that his name evokes in so many human minds?

Not, I think, in terms of Tarzan. It is time to approach the whole thing again right from the beginning: this time from the distaff side, and along a totally different route.

Two

The Escape Route

Once upon a time ... But which time? 'Man,' according
to the currently fashionable concept, 'is the child of the
Pleistocene.'

I am not going to begin with the Pleistocene. It was
a vivid and dramatic period when the climate of the
world went haywire and produced an era of prolific
evolutionary changes, and if you're talking about *homo
sapiens* there is no doubt that it was the Pleistocene which
first saw the finished product. The reason I'm not going
to talk about it is that the most fundamental ape-into-
man changes were already well under way before the
Pleistocene ever began.

The Pleistocene cannot answer the really difficult questions such as how and why our ancestors first began to
walk on their hind legs or how or why they first picked
up a stone and used it as a tool, for the simple reason
that these things happened before the Pleistocene ever
began. The Villafranchian (very early Pleistocene) hominids dug up in Olduvai Gorge were already walking
on their hind legs. They were already using tools. All
that was left to happen to them was an increase in cranial
capacity, an increased elegance in their gait, and the
acquisition of a chin. Before the Pleistocene came the
Pliocene, and before that again the Miocene, and we are
going to begin at the beginning.

Long long ago, then, back in the mild Miocene, there
was a generalized vegetarian prehominid hairy ape. She
had not yet developed the high-powered brain which today distinguishes woman from all other species.

A primatologist would have recognized her without much difficulty as a lively specimen of *Dryopithecus africanus* ('Proconsul') a primate who lived about that time and whose remains have been dug up in large numbers. Like present-day gorillas, she got her food from the trees and slept in the branches, but spent part of her time on the ground. But she was smaller and lighter than a gorilla; and she hadn't got the gorilla's confidence that her species could lick anybody in that neck of the woods. There were several larger species around that could frighten the daylights out of her.

After a couple of million years of this peaceful existence the first torrid heat waves of the Pliocene began to scorch the African continent. All around the edges of the forest the trees began to wither in the drought and were replaced by scrub and grassland. As the forest got smaller and smaller there wasn't enough room or enough food for all the apes it had once supported. The smaller and less aggressive species, and those least intolerant of descending to ground level, were driven out on to the open savannah, and she was one of them.

She knew at once she wasn't going to like it there. She had four hands better adapted for gripping than walking and she wasn't very fast on the ground. She was a fruit eater and as far as she could see there wasn't any fruit.

When frightened by a carnivore her instinct was to climb a tree or run away and hide, but there were no trees on the plain and very few places to hide. The man in the street will be surprised by her dilemma: he's seen films about prehistory and he knows all she had to do was pop into the nearest cave. But if you dropped him down at random somewhere in the middle of the veldt, he would be even more astonished to find that it's possible to wander around for weeks or months without ever seeing a cave to pop into.

In the forest she had often varied her fruitarian diet

by eating small insects and caterpillars and for a long time this was the only type of food she could find which looked to her remotely edible. She never thought of digging for roots—she wasn't very bright. She got thirsty, too, and the water holes were death traps with large cats lurking hopefully around them. She got horribly skinny and scruffy-looking.

You may be thinking that this eviction was just as bad for her brothers. Almost; but not quite. Remember, she was a primate. Primate babies are slow developers, and most primate females in the wild spend most of their adult lives either gestating or suckling a new infant or with the last one growing heavier and clinging on and slowing them down. It might be possible, in a transitional period between being vegetarian and learning to eat meat, to get by on a diet of grasshoppers; but if you're eating for two while this is happening you'll starve in half the time. Even before that, your milk dries up and the babies die.

For another thing, her brothers were probably stronger and better equipped. Her relative Proconsul, we are told, 'had large, fighting canine teeth'. Ardrey compares them to 'the magnificent daggers sported by apes and baboons'. But it just isn't true that baboons sport magnificent daggers. Only male baboons sport magnificent daggers. In vegetarian species these fangs are chiefly used for fierce intraspecies dominance-battles that females don't go in for very much. It may well have been the same for the Pliocene apes; so while her brother, when overtaken by something the size of, say, an ocelot, could inflict some very nasty lacerations, she could do little more than doggedly chew its ear while it unseamed her from top to bottom.

At this point people brought up on Tarzan will have a vague expectation that the father of her child will see what's expected of him, dash off and bring down an impala, drag it triumphantly back to her and say: 'There you are, darling. Help yourself.'

I'm sorry, but he'll do nothing of the kind. I've admitted she wasn't very bright—but he was just as thick as she was. Being unencumbered, he could get around faster, and like all primates he wasn't above sampling a piece of meat if it was brought to his attention. But if he happened on the remains of the lion's kill and managed to drive off the hyenas, it would never occur to him to give a piece of it to a female. Fruitarians have no need to develop these chivalrous instincts. On the contrary, if she happened to find a piece of offal on her own account he would promptly take it away from her. Ardrey rightly remarks about the dominant male that his 'instinctual objects of self-sacrifice seldom among primates include the female'. If they all looked likely to starve on that parching Pliocene savannah, he would make dead sure that she starved first.

In short, she found herself in an impossible situation. The only food in plentiful supply was grass, which her stomach wasn't designed to cope with. Everything in the vicinity (except the insects) was either larger, fiercer, or faster than she was. A lot of them were larger, fiercer, *and* faster.

The only thing she had going for her was the fact that she was one of a community, so that if they all ran away together a predator would be satisfied with catching the slowest and the rest would survive a little longer. This wasn't much of an advantage, though. If they all stayed together, pressure on the scarce and unfamiliar food resources would be greater than ever, with the females in their proper primate place at the end of the queue. The males, fresh from the trees, wouldn't have yet worked out the baboon strategy of posting fierce male outriders when the herd moved on; and if the predator always ate the slowest of the tribe, the cycle of gestation ensured that the time would soon come when the slowest would be you know who.

What, then, did she do? Did she take a crash course in walking erect, convince some male overnight that he must now be the breadwinner, and back him up by agreeing to go hairless and thus constituting an even more vulnerable and conspicuous target for any passing carnivore? Did she turn into the naked ape?

Of course, she did nothing of the kind. There simply wasn't time. In the circumstances there was only one thing she could possibly turn into, and she promptly did it. She turned into a leopard's dinner.

For her mate the impossible situation was just marginally less impossible. (That is why the predilection for the male pronoun has concealed the full hopelessness of their plight.) He lived a few years longer, but a chain is only as strong as its weakest link, and when he died no one came after him. Of course, the process took many more generations than this; it happened slowly over the torrid centuries as the forest dwindled, but the end was a foregone conclusion. They didn't have a cat in hell's chance. They became extinct.

At this point I anticipate a protest from the biologists and a yelp from the general reader. (I yelped myself when I first reasoned myself into this cul de sac.)

The specialist objection runs: 'It is simply untrue to imply that arboreal primates find it impossible to adapt to terrestial life. Baboons, mandrills, and macaques have done so with conspicuous success.'

Yes, it is certainly true that the baboons survived; so why not this little ape we are talking about? My own opinion is that the baboon's ancestors must have come to earth much sooner, and gradually learned to adapt to ground dwelling—by root digging, militant aggressiveness, social organization, etc.—under more propitious conditions, before the heat was on, because these adaptations take a long time to evolve. There are some solid anatomical reasons for believing that they didn't stay in the trees long enough to get as specialized for it as our own ancestors. For example, they never became brachiators, though most arboreal anthropoids begin to swing beneath the branches rather than running over them once they reach a critical size, and some of the early baboons, such as Simopithecus, were very large animals. We can be pretty sure that our own predecessors stayed aloft long enough to do a bit of brachiating, even though we never developed the elongated arms of gibbons and gorillas, because brachiating primates can move their arms in a sideways arc, through the crucifix position, while the baboon, like the dog, only moves his backward and forward.

The general objection is a more heartfelt one. If this primate who came down from the trees became extinct, what about the happy ending? *What about us?*

I will now come clean and admit she wasn't actually our grandmother, but a great-great-aunt on the maternal side who was unlucky enough to live in the middle of a continent. Hundreds of miles away near the coast lived a female cousin of the same species, another timid, hairy, undifferentiated Miocene-type ape. Her piece of forest was shrinking, too. As the heat and the dryness spread out from the baking heart of Africa, it became reduced to a narrowing strip; the larger and fiercer arboreans drove her away, just as her cousin had been driven, from poaching on their dwindling preserves.

She also couldn't digest grass; she also had a greedy and hectoring mate; she also lacked fighting canine teeth; she also was hampered by a clinging infant; and she also was chased by a carnivore and found there was no tree she could run up to escape. However, in front of her there was a large sheet of water. With piercing squeals of terror she ran straight into the sea. The carnivore was a species of cat and didn't like wetting his feet; and moreover, though he had twice her body weight, she was accustomed like most tree-dwellers to adopting an upright posture, even though she used four legs for locomotion. She was thus able to go farther into the water than he could without drowning. She went right in up to her neck and waited there clutching her baby until the cat got fed up with waiting and went back to the grasslands.

She, too, loathed getting her feet wet. It felt so unpleasant that she sometimes wished she had no fur at all. On the other hand, when your homeland's turning into an inferno the seaside's not at all a bad place to be. She found to her delight that almost everything on the beach and in the water was either smaller or slower or more timid than she was herself.

She switched easily, almost without noticing it, from eating small scuttling insects to eating small scuttling shrimps and baby crabs. There were thousands of seabirds nesting on the cliffs, and as she had a firm handgrip and a good head for heights she filled another empty ecological niche as an egg collector.

Besides the shrimps there were large creatures with harder shells, resembling mussels and oysters and lobsters. Her mate used to crunch through the shells or pry them open with his daggerlike teeth; she was envious of this because being daggerless she couldn't always manage it. One idle afternoon after a good deal of trial and error she picked up a pebble—this required no luck at all because the beach was covered with thousands of pebbles—and hit one of the shells with it, and the shell cracked. She tried it again, and it worked every time. So she became a tool user, and the male watched her and imitated her. (This doesn't mean that she was any smarter then he was—only that necessity is the mother of invention. Later his necessities, and therefore his inventiveness, outstripped hers.)

Whenever anything alarming happened on the landward side—or sometimes just because it was getting so hot—she would go back into the water, up to her waist, or even up to her neck. This meant, of course, that she had to walk upright on her two hind legs. It was slow and ungainly, especially at first, but it was absolutely essential if she wanted to keep her head above water. She isn't the only creature who has ever had to learn to do it. Although, as we have seen, she is almost unique in having learned to walk upright all the time, there is another mammal who does it for part of the time, and probably for the same reason. The beaver, whose ancestors also spent a good deal of time in shallow water, whenever she is transporting building materials or carrying her baby around, has the habit of getting up on her hind legs and proceeding by means of a perfectly serviceable bipedal gait.

She was very relieved to notice that even the large alarming-looking things that sometimes clambered out of the sea—things like seals, and giant turtles, and various kinds of sea cow, which were much commoner in those days—all proved to be very slow and clumsy and helpless on land, and in most cases totally disinclined to fight back when attacked.

Gradually her mate extended the shell-bashing manoeuvre to cover skull bashing as well. When you're dealing with dugongs or baby seals there is no risk involved and no call for beginner's luck or the accurate aim that a vegetarian would take centuries, or even millennia, to learn. You simply go on clobbering them with a pebble until they die, and then you eat them.

It wasn't very glamorous, but in the end he began to get quite a kick out of it. He learned to like the taste of meat as well as that of fish (seals and sea cows are both mammals) and became more efficient at killing things. Since they were both pretty well fed and there's an awful lot of meat on a sea cow, he didn't always make her wait until he'd finished before letting her have some. It took a few million years before he began to slide imperceptibly into the role of meat provider for the family, but there were so many alternative sources of food available to her that there was no particular hurry. Sometimes the carcasses got swept into the sea before they'd finished with them, so they took to dragging them up the beach and leaving them in a cave. It was the natural thing to do because the coastline is the place where you always find caves.

She spent so much time in the water that her fur became nothing but a nuisance to her. Oftener than not, mammals who return to the water and stay there long enough, especially in warm climates, lose their hair as a perfectly natural consequence. Wet fur on land is no use to anyone, and fur in the water tends to slow down your swimming. She began to turn into a naked ape for the same reason as the porpoise turned into a naked cetacean, the hippopotamus into a naked ungulate, the walrus into a naked pinniped, and the manatee into a naked sirenian. As her fur began to disappear she felt more and more comfortable in the water, and that is where she spent the Pliocene, patiently waiting for conditions in the interior to improve.

I believe these are the 'circumstances special to the point of disbelief' which explain how an anthropoid began to turn into a hominid. All the developments that otherwise appear strained, and improbable, and contrary to what we know of normal behaviour among primates and other quadrupeds, in these circumstances become not only credible and understandable, but natural and inevitable. Many features carelessly described as 'unique' in human beings are unique only among land mammals. For most of them, as we shall see, as soon as we begin to look at *aquatic* mammals, we shall find parallels galore.

Almost everyone has hitherto taken it for granted that

Australopitheca, since she was primitive and chinless and low-browed, was necessarily hairy, and the artists always depict her as a shaggy creature. I don't think there is any good reason for thinking this. Just as for a long time they 'assumed' the big brain came first, before the use of tools, so they still 'assume' that the hairlessness came last. If I had to visualize these early hominids, I'd say their skin was in all probability quite as smooth as our own.

However, we haven't reached Australopithecus yet, not by a long way. When I say the ape stayed in the water until conditions began to improve, I'm not just talking about a summer season. Suppose it took a couple of million years of drought to drive her into the sea; even then the African Pliocene didn't begin to let up for another ten million after that.

And a lot of strange and upsetting things can happen to a species in the course of ten million years.

Before we go on with the story, an acknowledgement is overdue.

This aquatic theory of human evolution was first suggested by the marine biologist Professor Sir Alister Hardy, F.R.S., in an article in *The New Scientist* in 1960. Later he gave a talk on it on the BBC's Third Programme, which was reprinted in the BBC's publication *The Listener*.

I heard nothing about it at the time. Apparently it made about as much impact on the scientific world as the reading of Darwin's first paper on the evolution of species to the Linnaean Society. (The president said in his annual report for 1858: 'This year has not been marked by any of those striking discoveries which at once revolutionise, so to speak, the department of science on which they bear. . . .')

Later Desmond Morris in *The Naked Ape* devoted a page or so to a full and fair summary of Professor Hardy's arguments, and acknowledged their 'most appealing in-direct evidence'; but something blocked him from going further. It may be that the traumatic experience of having almost drowned at the age of seven, which prevented his learning to swim for the next thirty years, prevented him also from accepting that we could ever have been bene-ficially moulded by so dangerous an element as water. For whatever reason, he dismissed the theory as unproven and, if true, of minor importance, a 'rather salutary christening ceremony'.

But I felt when I read that page as if the whole evo-lutionary landscape had been transformed by a blinding flash of light. I was astonished that after this key had been put into their hands, people were still going on writ-ing about the move from the trees to the plains as if nothing had happened.

Let's recapitulate a few more of Sir Alister's arguments. It wasn't only men's hairlessness that prompted him to suggest the idea. He remembered the fact noted and illustrated by Professor Wood Jones in his book *Man's Place Among the Mammals* that the vestigial hairs which remain on the human body—and which are seen still more clearly on a human foetus before it sheds its coat of hair—are arranged quite differently from the hairs of the other primates. It was Sir Alister who pin-pointed the exact nature of the difference—namely, that on the human body the vestigial hairs follow precisely the lines that would be followed by the flow of water over a swimming body. If the hair, for purposes of streamlining, had adapted itself to the direction of the current before it was finally discarded, this is precisely what we should expect to find.

He showed how the cracking open of shellfish would foster the use of tools. He pointed out that the ape was not the only mammal to arrive at this. Another aquatic animal, the sea otter, whenever he dives for a sea urchin, also brings up a stone in his other hand; he floats on his back, holds the stone on his belly, and uses it to break the shellfish on.

Professor Hardy wrote that wading in water would explain not only our erect walk, but the increased sensitivity of our fingertips, through the habit of groping underwater for objects we could not clearly see.

He pointed out that the best way of keeping warm in water is to develop a layer of subcutaneous fat, analogous to the whale's blubber, all over the surface of the body; that this is what all the aquatic animals have done; and that homo sapiens, alone among the primates, has in fact developed this layer, for which no other explanation has ever been found.

The more you think about it, the more impossible it becomes to believe that hunting man discarded his fur to enable himself to become cooler, and *at the same time* developed a layer of fat, the only possible effect of which would be to make him warmer.

Fur on the outside of the skin and a layer of fat beneath it both serve essentially the same purpose. The chief distinction between them is that one is better adapted to life on land, and the other to life in the water, and there is no conceivable evolutionary reason why any animal would begin to abandon one method in favour of the other unless its environment had undergone precisely this transformation.

The Hardy theory also explains why, however far from the sea they may be found, the very earliest man-made tools unearthed in connection with hominid remains are always fashioned from 'pebbles'.

Above all, it gives a simple and adequate explanation of the long chronological gap between the remains of Proconsul and his contemporaries, and the remains of Australopithecus. If no traces have been found of any creature transitional between the two, it could well be because the mortal remains of the naked apes and their first animal victims were not deposited in some Kenyan lair or midden, but swept out on the tide and devoured by fishes, while the first tools they ever chipped were mixed up with a million other pebbles like single straws in a haystack.

Since this theory was first mooted, some of the possible objections to it have been steadily undermined. For instance, some people found it hard to believe in that first plunge. Were not arboreal primates known to be averse to water?

Of most of them this was perfectly true. The anthropoids in particular were believed to fear water. In the wild they will not cross the narrowest rivers. They don't even need water holes, since they get enough moisture from their food and from the little pockets of rainwater that collect in leaves and tree trunks. It was a well-known 'fact' that chimps and gorillas are non-swimmers and any zoo could safely keep them in an unwalled enclosure by building a shallow moat around it. If by some accident they fell in, they would panic and drown.

But now hear Robert Golding, zoo curator at a Nigerian university, reporting on two gorillas aged six and a half and seven years old.

'The female in particular enjoyed having water hosed over her. When allowed access to the moat she went right into the water. The male was at first cautious but seeing her enjoying it, he followed. They now stand in the deepest part, up to their middles, and launch themselves forward in a sort of breast stroke. They do this many times a day. They seem to enjoy it—they make a noise, splash around, and play.... Seeing a man swimming on the other side of the barrier, the gorilla launches himself in a horizontal position with his arms straight ahead of him. It seems to come to him quite naturally.' It seems clear that, given enough incentive, our ancestors would have taken the same plunge.

Moreover, we have some reason to believe that they sometimes plunged pretty deep. In truly aquatic mammals, such as the seal, there is a special physiological development enabling them to dive and hold their breath for long periods underwater without running out of oxygen as quickly as a land mammal would. When a seal dives some of its metabolic processes slow down, reducing the body's consumption of oxygen. It can be measured most easily by the degree to which its heartbeat slows down. This physiological mechanism is known as bradycardia, and is found in many diving mammals, even fresh water ones such as beaver and coypu. It is also found in *homo sapiens*. When a man dives, his heartbeat slows down—not by any means as dramatically as a seal's, yet undoubtedly in human beings such a mechanism at some stage did at least begin to evolve. How, and when, and why? These things don't happen overnight, in the course of a holiday.

Some people rejected the aquatic theory because of the problem of the primate's baby, born so immature and helpless. Children three or four years old have been known to drown in a couple of feet of water. How could an aquatic ape ever survive the hazards of those first tender years? But now we read of Hollywood's water babies, film stars' tots dogpaddling merrily around swimming pools before they can even walk. Admittedly they are carefully coached by experts. What would happen if they weren't?

Anthony Storr provides us with the answer:

'The pioneer doctors who started the Peckham Health Centre discovered that quite tiny children could be safely left in the sloping shallow end of a swimming bath. Provided no adult interfered with them, they would teach themselves to swim, exploring the water gradually and never venturing beyond the point at which they began to feel unsafe.' If the prehominid's babies could do this, that Pliocene beach was the safest place for them in the whole of Africa.

The fact is that the Tarzanists, as well as forgetting the females, are constantly forgetting about the infants. It is many months before an anthropoid baby can be left alone. Its mother's existence is viable only because its fingers are strong enough almost from birth to cling on to her fur and so leave her four limbs free for going about her business. In such a perilous place as the open grass-land she would need that freedom more than ever; more than ever the infant would need not only its tight grip but something to cling to. The naked baby of a naked anthropoid would never have survived.

Only in the sea could the mother afford to dispense with her fur. The baby would have very few enemies in those four-foot shallows. Leopards don't come so far into the sea, nor sharks so near to the land. The child soon gets used to the water and once in he's mobile and comparatively weightless. All he needs by way of reassur-ance and support is to hang on, when he gets tired, to that part of his mother remaining above water, which is of course her scalp, so from that area of her skin the hair has never disappeared.

Professor Hardy explains the hair on our heads by say-ing that since only our heads remained above water, exposed to the sun, the hair remained to protect us from its rays. Other evolutionists, if they explain it at all, usually dump it on to the miscellaneous heap of unique human features labeled 'for sexual attraction'—a safe and lazy solution, since there are very few physical features which somebody at some time hasn't found sexually stimulating.

I feel that even protection against the sun is not a totally adequate explanation. If the hair was for this purpose it is true it would not have disappeared: it might have grown thick and tufty, as in many African tribes of course it is. But this theory leaves two things unexplained: the maiden with long flowing locks and the bald man.

In some populations of the ape there must have arisen, by mutation, the phenomenon of long hair on the head —a new departure for an ape. And the mutation must have proved adaptive. Why? I have seen this explained as a consequence of a move north or an Ice Age—protection against the cold. But this will not do. Cold is most acute when wind is blowing, and Jeannie's light brown hair, 'floating like a zephyr' on the breeze, would not have kept her body warm. The fact is that when monkeys from a warm climate are moved to northern zoos like Moscow's, they adapt by growing thicker hair all over their bodies. Climate might serve to explain the hairy Ainu, but not long tresses alone. Even for an aquatic ape, there must have been some advantage to outweigh the nuisance of its sometimes getting into the eyes, and taking so long to dry when its owner went ashore to sleep.

However, it would a powerful advantage for a baby if its mother's hair was long enough for his fingers to twine into; and if the hair floated around her for a yard or so on the surface he wouldn't have to make so accurate a beeline in swimming towards her when he wanted a rest. It would also explain the piece of dimorphism that nobody else has plausibly accounted for: in communities where the males took no part in the bringing up of the offspring there would be nothing to prevent their heads going as bald as their bodies, so long as this development remained sex-linked. Junior wouldn't mind Daddy's head being smooth and slippery because in the water, just as formerly in the trees, his mother was the one he hung on to.

There is one even more cogent reason for believing that the hair on a woman's head evolved for the benefit of her offspring rather than for the enticement of her mate. In the later stage of pregnancy it still happens that the proportion of thin hairs on the scalp becomes relatively smaller and the proportion of thick hairs relatively greater. The later stage of pregnancy is not a time when she has any particular reason to acquire extra sexual allure, and anyway the total visual effect is negligible. But as providing a safer temporary anchorage for a baby treading water, the development makes very good sense.

While we're on babies, let's take another look at breasts. A chimp suckles its young quite successfully with a pair of skinny little nipples located on a fairly flat pectoral surface, and there is no immediately apparent reason why the naked ape couldn't have done the same. But women are different; and the strongly favoured androcentric theory is that the difference is an aesthetic improvement, and that it evolved as some sort of sexual stimulus.

This is essentially a circular argument. 'I find this attribute sexy: therefore it must have evolved in order that I might find it sexy.' It's like saying that a woman walks with a wiggle because this is attractive to a male. In fact, she only walks with a wiggle because her children are so intelligent. The necessity of passing a large-skulled infant's head through her pelvic ring has prevented her skeleton from adapting to bipedalism quite as gracefully as that of her brothers; and males only find this defect attractive because they associate it with femininity.

Surely, if you are considering a process as strictly functional as lactation, and you notice a modification in the arrangements for it, it would be reasonable to think about the primary beneficiary of the process—namely, the baby —rather than trying to relate it to the child's father's occupation.

So—imagine now that you are this anthropoid baby.

You're having a whale of a time splashing around in the water, but after a while you get peckish. You pull your mother's hair and start bawling in her ear, so that she will come out of the water to feed you. A whale can squirt milk out to its pup rather like an aerosol container; but, as aquatic animals go, the whale is a pro and your mother is strictly in the beginners' class. Once or twice, being lazy or finding the rocks rather hard for sitting on, she's urged you to feed in the water, but there were waves, and your big brothers kept horsing around, and you swallowed great gulps of sea water and got terrible tummy upsets, so now she takes you ashore for the ten o'clock feed. She wades up the beach, sits up straight with water dripping out of her mermaid locks, holds you on her lap in the most natural position, with your head resting comfortably in the crook of her arm, and then relaxes and gazes absently out to sea, expecting you to get on with it as you and your kind have done from time immemorial.

But now, as the astronauts put it, you have a problem here. What the stupid woman fails to realize is that things have changed. There isn't any fur. If you let your head lie in the crook of her arm, the milk is high up out of reach. You have to hoist your torso into an erect position, and try to balance your head and somehow keep your lips clamped to this chimp-sized nipple of hers, and don't think it's easy. Your arms are too short to go around her waist, and if you scrabble around trying to get a purchase on something, there's nothing there but a faintly corrugated surface of slippery wet ribs. If she's a good type she will hold you up higher and help you, but she gets fed up with this much sooner because it takes more concentration and makes her arm ache, and any dairyman will tell you a milk producer won't give down properly if she's uncomfortable or irritated.

So you really need two things; you need the nipple

brought down quite a bit lower, and you need a lump of something less bony, something pliant and of a convenient size for small hands to grab hold of while you lie on her lap and guide your lips to the right place. Or, alternatively, guide the right place into your lips. And since you are what evolution is all about, what you need you ultimately get. You get two lovely pendulous dollopy breasts, as easy to hold on to as a bottle, and you're laughing.

Because of this new shape, and the fact that subcutaneous fat was being laid down all over her body at the time, a fair amount of this insulatory material naturally became concentrated in the breasts. And as Lila Leibowitz pointed out to the Northeastern Anthropological Association, the fat layer had other advantages—it cushioned the more fragile subtissue, it helped to keep the milk warm, and it stored reserve nutrients.

I don't think that in primitive conditions the form was typically hemispherical. In young females they would necessarily pass through a stage when they could be so described, and today, in civilized conditions, with high protein feeding, and school physical training, and sexual selection for the Adolescent Look, and birth control, and well-cut brassières, they may be coaxed into remaining that shape for quite a long time. But that's a form of neoteny—it's not the way they were originally designed, as any anthropological travelogue will amply confirm. Most men regard them as intrinsically hemispherical, but that's because whenever they imagine they are thinking about *Mulier Sapiens* what they are really thinking about is the Miss World contest.

So far so good. We have a possible explanation of the Raquel Welch phenomenon in this theory of the baby deprived of a handhold. It would be greatly strengthened if we could find an animal parallel, just as the shellfish/pebble-tool theory was strengthened by finding the sea otter. It would be nice to track down another mammal who went into the water, and found things happening to her vital statistics.

The trouble with aquatic animals is that some of them have been there so long that it's impossible to know where or how they lived before they went back to the sea. They've become as streamlined as fish. Nobody, for instance, can make a guess at the shape or habits of the unimaginable quadruped that lumbered down some prehistoric beach and began to turn into a whale (though we have some reason to believe it was actually quite small).

However, it is a fact that the only nonhuman pneumatically breasted females I have been able to trace happen to be aquatic. They are the Sirenians (or sea cows), that rare class of marine animals which include the dugong and the manatee, both credited with being the original 'mermaid'.

Each of them has been widely reported and believed to suckle while floating upright in the water holding its single offspring in its flippers. I haven't managed to find any reliable contemporary eye-witness account of this, but that may be because they are getting very rare, and their only close relative, the massive but inoffensive *Rhytina* —Steller's sea cow—was subjected to a campaign of systematic slaughter and is extinct. (Or let us say, since some vague rumours of a sighting drifted out of the Russian Arctic a few years back, almost certainly extinct.)

As to their statistics, the director of the Marine Biological Station at Al Ghardaqa describes the dugong as possessing a pair of 'well-developed pectoral breasts'. Steller wrote of the *Rhytina*: 'That they produce only one pup is concluded from the shortness of the teats and the number of the breasts'—which were two and pectoral.

The manatee is known colloquially in Guyana, according to David Attenborough, as the 'water-mamma'; and Colin Bertram writes of it: 'The breasts are indeed a single pair and pectoral, as in man.... In the manatee the teat seems to be almost on the actual hinder edge of the thick flipper just where it joins the body.' He points out that it would be impossible to tag a manatee by clipping a marker to the base of its flipper, as is done with seals, because the breast would be in the way; and he mentions that when the manatee is lactating the gland is 'large and shapely'.

So far the theory holds up. But is there any evidence that there was ever a time when they (and their offspring) had hands? I admit that the word 'manatee' has no connection with the Latin *manus*, a hand. But it is interesting to note that more than one keen observer, knowing more about zoology than etymology, has made the immediate assumption on looking at that jointed flat-nailed flipper that the creature must have been called manatee because of its hands.

The manatee's ancestor, of course, was nothing like a primate. It was certainly a land animal: it has the skeleton, the lungs, the vestigial hairs, to prove it. It is tempting to think of it as somewhat resembling the ancestor of its geographical neighbour the South American sloth, which must at one time have run along the branches before (like the orang) it grew too big and began to suspend itself underneath them instead. It is particularly tempting since the sloth wears her teats in the same eccentric position as the manatee—namely, under her armpits—and since the manatee and the two-toed sloth are the only two mammals in the whole of creation with six bones in their necks instead of seven.

But the taxonomists tell us that the sloth itself is not one of the sea cow's nearest living relatives. They are a small bizarrely assorted group and give us no help at all in trying to reconstruct a common ancestor. One is, improbably, the elephant. The second is a rabbitlike creature dwelling in holes in the rocks and referred to in the Bible as a 'cony'. The third and last is a small arboreal creature, the tree hyrax.

All we know for certain is that there must have been some ecological crisis (like the Pliocene for us) which induced the sea cow to leave her former habitat and take to the water: that the pectoral placing of the teats evolves most plausibly and most frequently in animals which at one time sat upright in trees: and that she has retained through all vicissitudes a vague instinct that her forelimbs were once for holding on with because she holds on to her infant with them, so that there might have been a time before she lost her fur, and when she still sat up on the beach to suckle him, when her infant likewise used his for holding on to her. If she did indeed leave the trees for the sea she is almost certainly the only creature besides ourselves that ever did so. Only instead of staying there for ten million years she stayed forever, and grew soggy and torpid, and lost her legs and most of the features of her face and degenerated into a great fat ugly six-foot blob of glup.

Poor cow, she's a far cry from Raquel Welch: one good look into those tiny watery eyes, and the mere thought that we might be sisters under the skin, would send most of us scuttling hastily back to Tarzan.

It would also make us wonder why on earth those jolly pig-tailed seagoing shanty-singing sailors ever took it into their heads to call her 'mermaid' and tell tall tales of her fatal magnetism. It can't have been only the rum ration: but we'll return to this problem a little later on.

Three

The Ape Remoulded

We are now beginning to build up a picture of the possible outward appearance of Mrs. *Australopithecus africanus* as she finally emerged after ten or twelve million years in the water.

We know that she stood and walked erect—or almost erect. We can tell from her skeleton that she couldn't lock her knees the way we can, so she probably stood with them slightly bent, and her gait would have seemed to us rather odd and ungraceful.

Sir Alister Hardy believes that she may have had webbed feet. 'In 1926,' he writes, 'Basler examined 1,000 schoolchildren and found that 9 per cent of the boys and 6·6 per cent of the girls had webbing between the second and third toes; and in some the webbing may extend between them all.' This development would not have been vital to the hominid if, as I am assuming, her habitat was littoral rather than marine and she didn't spend much time out of her depth; and as the aquatic phase was a very long time ago, it could have been almost entirely bred out again once it ceased to serve any useful purpose.

The fact remains that this phenomenon is unknown among other primates. But in case you were one of the 93·4 per cent of little girls who don't have any webbing between their toes, and if you are therefore sceptical of the whole story, try stretching your thumb and forefinger as far apart as they will go. Unlike an ape, you won't be able to make an angle much greater than 90 degrees; and what prevents you from increasing that angle is not the way your bones are arranged or articulated. It is that curious, functionless, vestigial, peculiarly human piece of thin skin in the angle of the thumb and forefinger.

If we had such a membrane under our arms which prevented us from raising them higher than shoulder level, we should seriously wonder whether we hadn't left the trees by planing gently to the ground like a flying phalanger, and whether Batman wasn't less an invention than a piece of race memory. I don't think the hominid's hands were in fact any more webbed than ours are, because it was always more important for them to be prehensile than to be adapted as paddles; but I do think it possible that that piece of interdigital skin was bestowed on us as a christening present.

So far, then, we have some reason to believe that by the time she was ready to emerge from the water she was a bipedal creature with smooth skin, limbs rounded by subcutaneous fat, and a thirty-four-inch bust or thereabouts. The skin was probably dusky, for the skin of most apes beneath the fur is already pigmented. The hair was probably long. If it was an aquatic adaptation, then long straight hair is conceivably the more primitive hominid pattern, and short frizzled hair a later, more sophisticated adaptation to life on land in hot climates, where heavy tresses would be a nuisance and serve no purpose. We'll come to the rest of her vital statistics later: first let's take a closer look into her face.

One facial feature distinguishing you very sharply from the rest of the apes is the shape of your nose. Monkeys' nostrils come in two main varieties, according to whether they come from the old world or the new, and this characteristic is used to name the two major groups of Catarrhine—old world, with a narrow partition between the nostrils—and Platyrrhine, the South American monkeys, with nostrils farther apart. Neither group has an arrangement in the very least like *homo sapiens*, who has taken the trouble to construct an elaborate cartilaginous roof over his nostrils and direct them neither forward nor to the side but straight downward, towards his feet.

Very little speculation has been expended on the possible reasons for this—probably for the Medawar reason that scientists don't ask questions very loudly until they can see the beginnings of an answer. For an aquatic ape the answer is perfectly obvious. If a gorilla attempted to dive or to swim under water, the water would be forced into her nostrils and driven under pressure into the nasal cavities and cause her the most acute discomfort. A seal avoids this by having nostrils which it can open and close at will. The aquatic ape avoided it just as effectively by modifying the shape of her face so that the water would be deflected by a splendid new streamlined structure and her sinuses would be safe.

It may or may not be pure coincidence that the other primate who has gone to some lengths to cover his nose with a lid is the only one apart from ourselves who regularly takes to the water for the sheer joy of it. It is the proboscis monkey, a Borneo species of langur. Irven de Vore says: 'These odd monkeys delight in swimming; a favourite escape from midday heat is to drop into a stream and thrash about with a crude dogpaddle,' and their faces are a rather crude caricature of the face of Schnozzle Durante.

The next question is, When did the human nose arrive? for it was presumably after we too began to 'delight in swimming'. And the answer is: very early indeed. A comparison of a side view of the skull of a chimpanzee and that of the early hominids shows that while the chimp's face is concave from brow to jaw, the precursors of homo sapiens show a modification in this structure, a bony projection to which nasal cartilage could be attracted. W. E. le Gros Clark says: 'The nose of homo erectus was broad and flat, as it is in certain races of mankind today.' The significant fact is that it was there at all—yet another factor easy to explain in terms of aquatic adaptation, very difficult to explain by any other means, and arising like the erect posture and the use of tools long before *homo* ever became *sapiens*.

So our hominid had a nose. I have no doubt that she also had fleshy nostrils, but considerable doubt that they evolved to make sex sexier for her mate. I think she was by no means the simian, cadaverous, lipless creature that artists sometimes reconstruct by covering her dug-up skull with a tightly fitting layer of hairy skin. The layer of fat which was rounding out her arms and legs and adding bulk to her breasts was also filling out her cheeks, and her nostrils, and her earlobes, and everting her lips. Even today on a very thin person, or one whose subcutaneous fat has wasted away with age, the lips are sometimes barely visible, but on a plump one they are always a noticeable feature. We would not have accounted her beautiful, with her low forehead and prognathous jaw, but the chances are that she was a chubby little creature with several superficial features resembling our own more nearly than they resembled any ape's. And as for the expressions that flitted across that prehistoric countenance, her millions of years in the water had certainly left their mark on those also.

Charles Darwin wrote a whole fascinating book about *The Expression of the Emotions in Man and Animals*. In comparing man with our primates, he found enough similarities to reinforce his conviction that they shared a common ancestry.

Man, apes, and monkeys can all be observed to cry out when in pain, flush when enraged, yawn when tired, glare when defiant, grin when tickled, tremble when afraid. embrace when affectionate, bare their teeth when hostile, raise their eyebrows when surprised, and turn their heads away when offended.

But the expressions that caused Darwin most trouble were the ones that man alone has evolved, especially when inquiries to the more distant parts of the world confirmed that several of these expressions were not merely European cultural conventions, but common to the whole species. 'Men of all races,' Darwin observed, 'frown.'

By this he did not mean the steady glare from under lowered brows by which the apes sometimes convey displeasure. He meant such features as the wrinkling of the brow, and the 'obliquity of the eyebrows' brought about by what he called, 'for the sake of brevity, the grief muscles'. The eyebrows, he commented, 'may be seen to assume an oblique position in persons suffering from deep rejection or anxiety; for instance, I have observed this movement in a mother while speaking about her sick son; and it is sometimes excited by quite trifling or momentary causes of real or pretended distress.'

Now, we certainly didn't inherit these expressions from the tree-dwelling apes. Darwin points out:

'In comparison with man, their faces are inexpressive, chiefly owing to their not frowning under any emotion of the mind—that is, as far as I have been able to observe, and I carefully attended to this point.... I made my hands into a sort of cage and, placing some tempting fruit within, allowed both a young orang and chimpanzee to try their utmost to get it out; but although they grew rather cross, they showed not a trace of a frown. Nor was there any frown when they were enraged. Twice I took two chimpanzees from their rather dark room suddenly into bright sunshine, which would certainly have caused us to frown; they blinked and winked their eyes, but only once did I see a very slight frown. I have never seen a frown on the forehead of the orang.'

Indeed, the apes make so little use of the corrugator muscle that causes frowning that Sir C. Bell at that time believed it was peculiar to man and called it 'the most remarkable muscle in the human face'. The same distinction can be made about what Darwin called the lower orbicular muscles, which wrinkle and compress the lower eyelids and give unique expressiveness to human smiles, tears, and laughter.

Darwin believed that the main reason for frowning and allied expressions such as grief—and possibly also the trigger that turns on human tears—is the fact that in a child the emotions that induce frowning might if intensified induce screaming; and in loud screaming the eyes might become engorged with blood unless we protected them against this danger by tightly closing the eyelids or otherwise compressing the eyeballs by the muscular exertion of frowning.

I think the explanation of these differences between ourselves and the apes is much simpler than this. I think it is simply that the forest-dwelling apes lived in a dim and dappled land with a screen of leaves over their heads; that they never adequately developed the muscle which would enable them to frown at a bright light because the bright light seldom if ever shone on them.

Now consider our aquatic ancestor, with nothing over her head but sky, and the sun burning down through all those cloudless Pliocene millennia, reflecting back up from the surface of the water, shining and splintering and sparkling and dazzling into her unaccustomed pongid eyes.

If the baby was swimming on the sunward side of her she'd have some difficulty in seeing him at all against all that brilliance; and still more in distinguishing her own child from another. Today, on such a coast, she'd invest in sunglasses. All she could do then was try her hardest to bring her eyebrows farther down and closer together and cut down some of the glare from above, and contract the lower orbicular muscle against the reflections from below. Frowning, then, became a built-in response to difficulty, obstruction, bafflement.

As for her mate, he had similar problems. There weren't many hostile species around, but a male primate is generally much bothered about dominance; he thinks any member of his own kind might turn into an enemy if he doesn't watch out. When anyone approached him he would stare across the water, contracting the skin all around his eyes to reduce the dazzle, trying to get a good look at who it was and what his intentions were and whether he was beginning to act uppity. So frowning became a sign of hostility and anger also.

Finally, think about that baby, whom we must never forget. Whenever he felt frightened or anxious or sorry for himself he tried to look into his mother's face to see if she knew something was wrong and whether she intended doing something about it. Neither his father nor his mother had much occasion to look up, so his job was really the toughest of all. He had to try to bring his eyebrows in from the sides to cut out the excess glare, while not letting the middle bits of them come down to obstruct the view of his mother's face directly above him. Darwin describes his final solution perfectly: '... the contraction of certain muscles which tend to lower the eyebrows is partially checked by the more powerful action of the central fasciae of the frontal muscle, which raises the inner ends alone....'

Over the millennia the baby perfected this expression; until even after he'd grown up his face often fell instinctively into that shape when he was feeling distress of body or mind and knew that cause of it was something he was powerless to cope with on his own. Our faces fall into it, too, when we are distressed and helpless, and we recognize it immediately in others as a sign of grief. It is exclusively human.

Darwin said of it: 'During several years no expression

seemed to me so utterly perplexing as this we are here considering.' He even remarked on the strikingly similar effect elicited when he 'made three of my children, without giving them any clue to my object, look as long and as attentively as they could at the extremely bright sky.... The eyebrows and forehead were acted on in precisely the same manner, in every characteristic detail, as under the influence of grief or anxiety.'

In spite of this he returned in the end to his theory of defending the eyes against the danger of becoming engorged and suffering damage in a possible screaming fit. He did in fact refer to the need of protection against overbright light, but only as a secondary contributing factor. I think if Sir Alister Hardy had been there to freshen his thoughts with a splash of sea water, he would perhaps have gone all the way and opted for the simple explanation instead of the complex one.

Wait a minute, though: I'm jumping the gun in calling it sea water. Although we have plenty of evidence of immersion, we have nothing so far that would establish whether we were immersed in salt water or fresh. For the hippo lost its hair in fresh water; the beaver learned its bipedal shuffle in fresh water; the manatee's preferred habitat is estuary and river water, and it sometimes follows a river upcountry for hundreds of miles.

Is it possible that we were lake dwellers rather than sea dwellers? Do we have any hope, all these millions of years later, of finding any clue at all to the actual nature of the water those aquatic apes were swimming in? Well, there's no harm in trying.

Sea water has one major drawback: unless you're a fish you can't use it for drinking. A mammal the size of a woman has an urgent and imperative need to absorb at least 500 cc.'s of 'free' water per day to keep its kidneys working (not counting what it needs for perspiration). If its kidneys can't keep working it will very soon die; and for this purpose sea drinking water is useless, because the osmotic concentration of human urine is hardly any higher than that of sea water.

This applies to practically all mammals. Even the camel, in spite of legend, couldn't possibly use sea water for drinking and survive on it. In fact, there is only one exception to the rule—a little kangaroo rat called Dipodomys, who lives in Arizona deserts so ferociously arid that he can't afford to sweat at all, and he can concentrate his urine to osmotic levels over four times as high as ours—so high in fact that after leaving his bladder it is apt to solidify. He could drink sea water without any ill effects at all—he's done it, in laboratories—and it's just one of life's little ironies that out in the Arizona desert he never manages to find any.

Anyway, our hominid was no Dipodomys. She was more like the Ancient Mariner—water, water, everywhere, nor any drop to drink. Perhaps, like the manatees, she hung around estuaries, where what was left of Africa's rivers found their way to the sea. Perhaps she went ashore in the early morning and looked for plants with the dew on them. She could keep down her need of free water by saving 1000 cc.'s or so per day on perspiration, through staying in the water by day and sleeping in a cool cave at night. The fact remains that if most of her food came out of the sea, a certain amount of salt water must have found its way into her interior and wasn't doing her very much good.

How do other marine creatures deal with this problem? Some of the mammals have adaptive mechanisms for keeping the sea water out of their stomachs: the whalebone whale has a giant sieve in its mouth for straining out a huge mouthful of plankton and small fish, which it then presses against its palate with its tongue into a solid mass of pulp; and when the harbour seal swallows a fish under water, it has a specially designed esophagus that wipes the fish virtually dry on its way down. Their requirements of water are met by the moisture contained in their food. They don't sweat, and can at need reduce the renal blood flow and filtration rate to very low levels.

More interesting still are the sea birds. A bird's kidney is no more able to cope with the ingestion of sea water than ours is; yet birds like the shearwater, the petrel, and the albatross spend months on end far out of sight of land. In 1956 Knut Schmidt-Nielsen carried out an investigation into the salt and water balance of the double-crested cormorant. Homer W. Smith recounts:

'To determine what would happen if sea water were ingested, quantities amounting to about 6 per cent of the body weight were administered by stomach tube. As was to be expected, the concentration and rate of excretion of urine were quickly increased, chiefly in relation to increased excretion of sodium chloride. But what came as a complete surprise was the secretion of clear waterlike liquid by two glands in the head which drain into the internal nares and are known to anatomists as "nasal glands". This liquid ran from the nasal openings and down the beak to accumulate at the tip from which the drops were shaken off by sudden jerks of the head. The secretion proved to be an almost pure solution of sodium chloride.' That bird was weeping salt tears.

Anatomists had known about those glands for years. They knew that in marine birds the glands were greatly enlarged, with a richer arterial blood supply and more highly developed glandular structure than in terrestial forms. They knew that even within a single genus such as the gull, the size of the gland increases with the extremeness of the marine habitat. But until 1957 they never knew why.

Once this discovery had been made, they took another look at marine species other than birds. They investigated the marine terrapin and discovered that if he swallowed salt water he wept salt tears; they proved the same thing about the marine loggerhead turtle.

Everyone knows of the old legend that crocodiles weep crocodile tears. Until recently most of us 'knew' that this was merely a fable; that men who had known and worked with crocodiles for years were prepared to swear that they had never once seen it happen. Now we know that the non-weeping crocodiles are fresh-water crocodiles, and the anatomy of the nasal glands of the salt-water crocodile makes it clear that if he got a bellyful of sea he would certainly cry his eyes out.

Do you want any more? There are land snakes, fresh-water snakes, and sea snakes. There are land lizards, and at least one sea lizard, the marine iguana of the Galapagos. In all cases, the marine forms have the paired nasal glands highly developed; and whenever it has been possible to capture a specimen and make it swallow sea water, the result has been the same.

Now, we are neither birds nor reptiles. The glands from which our tears flow are not directly comparable to the glands from which their tears flow—but then even among themselves the arrangements differ, and the only thing they all have in common is that from somewhere or other in the vicinity of eyes or nose or beak a saline liquid emerges and flows.

Any scientist will tell you that our tears will not serve the purpose of the petrel's tears because they are not sufficiently concentrated. True. Like the webbing between the toes and the thickening of the pregnant woman's hair and the slowing of the heart in a dive, it was only ever, probably, an incipient adaptation. We never got very far with it because we didn't remain aquatic long enough; but it had begun.

The ineluctable fact is that we are the only weeping primate of any kind. And if you want to find the only weeping carnivorous mammals, you have to go into the sea again, and meet the weeping seals, and the weeping sea otters.

You may object that we don't cry because we have swallowed sea water; we cry when we are emotionally upset. This is quite true of us: it is true also of all our marine comrades.

Homer Smith records of the albatross: 'Nasal dripping was observed to occur when the birds had been fighting with each other, during their ritual dancing, or even during the excitement of feeding time.' He concluded that the nervous control of the gland was liable to react in this way during moments of stress.

R. M. Lockley wrote in his book *Grey Seal, Common Seal* that the seal's tears 'flow copiously, as in man, when the seal is alarmed, frightened, or otherwise emotionally agitated'.

Steller wrote of the sea otter: 'I have sometimes deprived females of their young on purpose, sparing the mothers themselves, and they would weep over their affliction just like human beings.'

Up to now no one, not even Darwin, has offered a convincing explanation of the origin and purpose of human tears. I would suggest that since the only weeping birds are marine birds, and the only weeping crocodiles, snakes, lizards, turtles, and mammals are marine crocodiles, marine snakes, marine lizards, marine turtles, and marine mammals, it is surely not beyond the bounds of reason to suppose that the only weeping primate was once a marine primate.

I believe that the water the aquatic ape splashed around in was as salt as the tears she shed when her new grief muscle came into play. And pretty soon she had plenty to grieve about; for in a different part of her anatomy even queerer things were happening to her.

At one point in *African Genesis*, Robert Ardrey urges

you, the reader, to go into the bathroom, lock the door, undress, and gaze into the mirror at the primitive unspecialized nature of your anatomy. He points out that you will note certain characteristics distinguishing you sharply from the other primates—hairless body, large head, a chin, flat feet, large buttocks, and, compared to the apes, short arms and a small chest. (He doesn't expect you to notice anything else unapelike about your chest because he naturally thinks his reader is a man, and he doesn't mention your nose because he doesn't know why it's there.)

Two of these developments in particular, he says, would enthrall a visiting zoologist from a distant planet because they are the truly human specializations—the flat feet, and the 'thoroughly outsized buttocks', with its specially developed mass of muscle permitting turning, twisting, and balancing in an erect position.

When it comes to accounting for this muscular mass, there is the usual division of opinion. Sir Alister Hardy would maintain that swimming, diving, turning, and manoeuvring in a three-dimensional element such as water would naturally encourage great flexibility in the spine, and a muscular plexus to facilitate these new movements would inevitably evolve. The Tarzanists believe that it emerged in response to the hunter's need of throwing missiles and spears.

As a matter of fact, the otter is very good at throwing things, while its terrestial relatives—stoats, pine martens, etc.—show no signs of this ability; which suggests that when the hunter hurled his first spear an aquatic background, if he had one, would certainly have helped. However, what I would like to examine now is not so much the cause as the consequence of this new anatomical structure, the buttocks.

In the average pronograde quadruped, the rear end of the body is not exposed to a great deal of wear and tear.

It's a protected area. Any hostile encounters are met head on; the weight of the body during waking hours is sustained by the legs and feet, and during sleep in most cases by the flanks. Even in animals like dogs and cats, which 'sit' erect, the limbs are so designed that in this sitting-up position the weight is still distributed between the long heel bones of the hind limbs and the toes of the front ones.

In this carefully protected posterior area are located certain important and vulnerable orifices, such as the anus, the urethra, and the vagina. In the great majority of mammals they are afforded extra security by the addition of a movable tail which neatly covers and protects them, when they are not in use, against cold and rain and accidental damage. It's a very neat and efficient arrangement.

At least for a ground-dwelling mammal it's an efficient arrangement. Even for a tree-dwelling primate it's an efficient arrangement if you are a small one, like a tree shrew, and can curl up and go to sleep on a big fat branch, like a cat, or creep into a hole in a tree trunk.

Many of the primates, however, didn't keep to this small and convenient size. They grew so big and heavy that it would have been perilous or impossible to curl up and go to sleep on a branch. Even when awake they couldn't feel as relaxed lying along a branch as a panther or leopard can, because their limbs were differently specialized for reaching out and picking things, and wouldn't fold up tidily underneath them. The way they felt safest, when at rest, was when they had at least one hand holding on to a good sturdy branch somewhere around the level of their heads.

They found the most secure and comfortable way to relax was sitting up in the crook of a branch—just where it joined the trunk, so that it wouldn't sway in the wind or snap off.

This meant that for the first time their bottoms were taking the weight, which they'd never been designed to do. They got very sore, as you can imagine, because they were much bonier bottoms than ours. After putting up with this for some time, they invested in a pair of leathery protective patches known to primatologists as 'sitting pads'. (In some old world monkeys this protective adaptation has affected not only the skin but the skeleton, resulting in modifications of the pelvic bones known as 'ischial callosities'; but we can be pretty sure that our own ancestors contented themselves with the more superficial version.) These ensured that their orifices were still okay, because the crooks of trees they sat in were V-shaped; the steadying handhold took part of the weight, the sitting pads protected them against friction, and the orifices encountered nothing but the air. In that way it was no more uncomfortable for them than sitting on the lavatory.

Some of the bigger arboreal primates (including our own ancestors) found that by the time they'd reached this stage the tail was a bit pointless and they discarded it altogether. The very biggest ones, like the gorilla and the orang, finally grew too big and heavy to squash into the crook of a branch at all without the greatest discomfort, and they took to building nests out of branches, as the only way of ensuring a good night's sleep.

We can't be quite sure at what stage of this process our prehominid went into the water. Certainly she'd lost her tail. I don't believe she was big enough to have started nest building; because animals that revert to the water have a tendency to grow bigger, and even the Australopithecines that emerged millions of years later weren't particularly big.

It's fairly safe to assume that in her arboreal days, when she wasn't running along the branches or venturing down to the ground to explore, she was sitting up in the branches, to eat, to look about her, to nurse her baby, or to sleep. And in the trees it was perfectly tolerable.

But sitting on the beach was a very different matter. A quadruped's bottom isn't at all like ours. When she first took to a littoral life she had nothing in the way of padding there. Her vagina was in the normal quadruped position, just under where her tail would have been if she had one; it was normal also in being exposed, flush with the surface for easy access. Sitting there on the pebbles and the salty shingle and the wet sand and the rocks and barnacles, with a growing anthropoid infant on her lap, must have been hell.

Fortunately, this stage didn't last too long, because the marine environment at once set in train two separate morphological developments, each of which proved (initially, at least) benign.

The first, as we've seen, was the marine mammal's layer of subcutaneous fat; at the same time as she was using some of it to keep the baby happy, you can bet she was laying down also a pair of posterior hemispheres as fast as evolution could accommodate her. And I think (*pace* the androcentrics) that however much pleasure the sight of them may have afforded her mate, she evolved them chiefly for her own convenience and out of her own dire need.

There were muscles there, too, as we have noted, and for whatever reason they evolved they added to the solidity of the new structure. Just how impressive it was we cannot be quite sure, for a behind is a perishable feature that doesn't show up on Austrapolithecine skeletons. Among the Bushmen, the most ancient people of Africa, we find the phenomenon of steatopygia, the accumulation of very large deposits of fat on the buttocks—they may even project to the point where 'you can balance a wineglass on them'. It is impossible to know whether this was an offbeat development or a formerly common feature later discarded.

What we can observe, even among ourselves, is a certain amount of sexual dimorphism in this respect The male has somewhat less of a bottom than the female. Any attempt to account for this is usually avoided by dropping it into the large miscellaneous grab bag labelled 'for sexual attraction'; but there may well have been real reasons for it in the beginning. If it evolved, as the ischial callosities had done, as a protective feature, then the female ape would need it more. The male had only one orifice there instead of three, and that was a rather less sensitive one; besides, he didn't have to spend time sitting around holding the baby, which as long as it was breast-fed she would have had to do.

Simultaneously another change was taking place in the female, which was primarily a by-product of her new upright mode of locomotion. When she began to stand upright, the normal quadruped 90-degree angle between her spine and her hind legs had become 180-degree and this had displaced some of her abdominal internal organs, resulting in the change noted by Desmond Morris:

'There is the basic anatomy of the female vaginal passage, the angle of which has swung forward to a marked degree, when compared with other species of primates. It has moved forward more than would be expected simply as a passive result of becoming a vertical species.'

It not only moved forward—it also withdrew farther within the body cavity, possibly for additional protection against salt water and abrasive sand. This is a normal marine modification. The ears of seals—except the otaries —became interiorized by a similar process, for purposes of protection and streamlining; and their nipples became retractable and covered by a flap of skin.

It is natural to expect that the aquatic ape would follow as far as possible this pattern of internalizing exterior organs wherever practicable and in addition protecting them with a covering membrane wherever practicable. She did this to her vagina too—she not only retracted it but covered it with a protective membrane known as the hymen. It was not practicable for this to remain intact all her life, but it was better to have it for ten or twelve years than never to have it at all.

It is of course a prime tenet of all andocentric thinking that everything about the female was designed primarily for the benefit and convenience of the male, to make her (a) more attractive to him and (b) more accessible to him; and if ever you want a really good laugh, I would recommend reading up some of the incredibly involved and convoluted arguments of a male evolutionist trying to explain why woman, alone of all the apes, equipped herself with a hymen, which appears on the face of it to have no other purpose than to keep him out.

Now at last, with her vagina tidily tucked away and a long way forward of the weight-bearing area of her bottom, and a pair of well-cushioned buttocks as well, the hominid was sitting comfortably, and might have imagined that the worst of her troubles were over.

Alas, they were only beginning. Her mate was by this time regarding her with an increasingly jaundiced eye

Look at it from his angle. The normal procedure for primate mating is simple. The male mounts the female from behind. She braces herself to take the weight on all four limbs, and if she has a tail she keeps it well out of the way. A monkey's usual technique is to stand on the joints of her back legs with his feet. If the female is uncooperative he just can't make it at all, but if she is willing—and oftener than not she is even more eager than he is—he encounters no difficulty whatever. Compared to ours, her legs are spidery and handsomely arched like a croquet hoop, and her vagina is a surface organ located for maximum convenience This was what the male hominid's ancestors considered themselves entitled to expect.

So no doubt did he. But he wasn't getting it. For one thing she was no longer accustomed to supporting even her own weight on all four limbs; they were developing a disconcerting tendency to buckle at the elbows and knees and subside. This alone wouldn't have worried him —other primates have adapted to this sort of thing.

But at the same time her legs, instead of being slender and apart, were growing as gross as tree trunks from the knees up, and so close together that when she was standing still he very often couldn't see daylight through them. He wondered gloomily where it was all going to end, and whether she intended ultimately to fuse herself into a single streamlined column like a walrus or a sea elephant. And the curves again—he felt she was carrying them a little too far. It was all very nice and pneumatic for her, carrying her own air cushion around with her, but for a rear-mounting male it simply made things unnecessarily difficult.

Her own demeanour was as naïve as ever—she would romp up to him full of high spirits and lubricity and present her buttocks as of yore. But you have to remember that when all this started he was by no means *homo sapiens*, with a larger penis than any other living primate. He more nearly resembled the chimp, a far less well endowed creature. With every million years that passed, with every centimetre of the vagina's steady progress forward and inward, he came more and more to feel, as his little son had felt in a different connection: 'It's all very well for the stupid bitch to carry on like that. What she fails to realize is, things have changed. I've got a problem here.'

As you see, we now have new and revolutionary answers to two of the vital questions propounded by Desmond Morris. The new answers are less ingenious, less elaborate, and I'm afraid somewhat earthier, but like all the aquatic answers they have the merit of extreme simplicity.

Question one: Why did homo sapiens develop the largest penis of all living primates? Not because he was a hunter and had to keep his pack happy by pair-bonding and had to cement the pair bonds by making sex sexier. It grew longer for the same reason as the giraffe's neck— to enable it to reach something otherwise inaccessible.

Question two: Why did he switch from rear-mounting to a frontal approach? Not because the incoming signals from his beloved's lips and eyes would make the experience a more memorable personal affair and thus help to induce monogamy. He came around to the front because he could no longer make it from the back.

Once again, if you are asked to believe that man's sexual approach has any possible connection with an aquatic phase in his history you might at first find it hard to swallow. But once you begin to realize that practically *all* land mammals use the rear approach to sex, and practically *all* aquatic mammals use the frontal or ventroventral approach, then you are bound to suspect that the connection must be more than fortuitous.

There is almost no terrestial mammal whose sexual modus operandi differs basically from that of the cat or the dog or the monkey. (I'll grant you the porcupine, but his motives are obvious.)

For the sake of completeness, I'll offer one more exception. (As with Dipodomy's kidneys, there's always an exception if you dig around hard enough.) In this case the exception is nearer home—a primate, and an anthropoid. Orangutans are by no means wedded to the rear-mounting procedure. But the reasons for this are fairly evident and practically unique.

They are such confirmed brachiators that of their own choice they never leave the trees. In their natural habitat they spend most of their working lives suspended from the branches either by their hands or by their feet or both, and they are extremely large and weighty animals. In these circumstances it is generally quite impracticable for the female to balance quadrupedally on a branch to facilitate rear mounting; so they practice an end-to-end technique during which both can remain suspended. In captivity, where there are no branches, it is not unusual for them to copulate with one or both animals lying on their backs. Sloths, the other permanently suspended species, must almost certainly have evolved a similar solution, though I have read no account of this; and the gorilla, the only other heavyweight brachiator, may have passed through a stage of at least experimenting with this method, for although now he spends most of his time on the ground and prefers the normal quadruped rear approach, he has occasionally been observed in zoos using an orang-type method—female supine, male sitting up.

On the other hand, once you begin to look into the sea, the ventro-ventral approach is not an aberration: it is practically obligatory.

Of course, we cannot possibly maintain that all animals returning to the water adopted at some time or other an erect posture. What we can say is that all the ones who became fully marine have turned into streamlined swimmers; and a streamlined swimmer, just as much as a walking biped, finds the 90-degree angle between spine and hind limbs extended to 180 degrees, with consequent displacement of internal organs and the probability that in the female the vagina will tend to migrate to the ventral side. This is what we would expect to happen, and this is what has happened.

Steller wrote this eyewitness account of the *Rhytina*:

'In the spring they mate like human beings and especially towards the evening, if the sea is calm. Before they come together many love games take place. The female swims slowly to and fro, the male following. He deceives the female by many twists and turns and devious courses until finally she herself becomes bored and gets tired and forced to lie on her back, whereupon the male comes raging towards her to satisfy his ardour and both embrace each other.'

Victor Schaffer describes the mating of the largest living mammals in *The Year of the Whale*:

'Hour after hour the pair swim side by side, keeping touch by flippers and flukes, or simply rubbing sides.... Presently the male moves to a position above the female, gently stroking her back.... The Cow turns responsively upside down and the bull swims across her inflamed belly.... At last the pair rise high from the sea, black snouts against the sky, belly to belly, flippers touching, water draining from the warm, clean flanks. They copulate in seconds, then fall heavily into the sea with a resounding splash.'

Colin Bertram in his book *In Search of Mermaids* quotes this account of the behaviour of manatees as related by an eye-witness in 1955:

'The manatees were disporting themselves in the river towards the left bank and in a school of fourteen to sixteen; they gave the impression of fighting among themselves. Later they moved into the shallow and worked themselves up the bank into six inches of water. One pair was completely out of the water. They mated lying on their sides.' (As the organs are ventral, this must have meant face to face.)

Since the introduction of sea aquariums we are becoming much more familiar with porpoises and dolphins. When I visited one and asked the trainer how dolphins mate, he placed his hands together palm to palm and said: 'Like that.'

Seals and their kind are the only notable exception. In the case of seals, sea lions, etc. the favoured method is still rear mounting, and the sex organs have remained near the tip of their tapering bodies. The reason for this may be that they have clung to the habit of coming out on to the rocks to copulate. You can see their dilemma most clearly if you consider the case of the sea elephant. The male may be up to two and half times the weight of the female and weigh a couple of tons. He heaves himself around like a juggernaut, and hundreds of infant sea elephants die every year because the males are too clumsy to avoid them and they are simply crushed to death. If the female offered herself supine to such a suitor instead of with her spine to the sky, her internal organs would probably never be the same again.

Of course, when elephant seals copulate underwater, as they sometimes do, the ungainliness vanishes, and they make love with the airy grace of ballet dancers; and at least some species of seal under these conditions adopt the more conventional marine-mammal approach. Gavin Maxwell describes the behaviour of seals in the Shetland Islands before the breeding season proper: pairs of seals 'roll and twist in the water, snarling and snapping at each other.... This play is thought to be equivalent to courting, as copulation appears to occur afterwards, the cow rolling on her back with the male on top gripping her with his flippers.'

Indeed, face to face for aquatics is so consistent a rule that they don't even have to be marine to resort to it—it holds good for fresh water divers also. The actual mating of beavers is not often seen, but it has been observed on Russian beaver farms, and, as Lars Wilsson relates it: 'The scent of the female in heat is presumably enough to make the male sufficiently sensually stimulated, and when she goes into the water in a particular way he follows, and mating takes place stomach to stomach, the animals swimming slowly forward.'

And to return to the sirenians, a report on the dugong by H. A. F. Goohar confirms that these tropical sea cows fit into the face-to-face aquatic pattern. The same report offers the most probable solution to the mystery of the mariner and the mermaid. It points out that there is a striking resemblance between the genitalia of dugongs and those of human beings; and that in the Red Sea area there is an oral tradition that in former centuries a sailor after months at sea who found a dugong in the shallows—large, docile, warm-blooded, air-breathing, smooth-skinned, female-breasted, and with ventral genital organs which remarkably well fitted his own—wouldn't worry overmuch if she was comparatively faceless. In those days such a sea maiden may well have represented temptation even though she has never learned to sing a siren's song.

I submit that the major change in our sexual behaviour, and the major modifications in our physical sexual structure differentiating us from all other primates, constitute evidence as clear as any other that at one stage of the game we suffered a sea-change. The reasons for the change are not too difficult to deduce; and the resulting modifications, which can be paralleled in so many species of aquatic animals, can be paralleled nowhere else in the world.

If we had followed their example and forsaken the land altogether we would never have become the lords of creation, but neither would we be the crazy mixed-up creatures we are today. If I may for a moment be thoroughly anthropomorphic, most of the species that go back to the water seem to get no end of fun out of life.

Penguins appear to be at once more placid and more playful than birds. Otters seem to be twice as intelligent, three times as curious, four times as friendly, and ten times as lighthearted as their land-bound relatives the weasels and the ferrets and their kind. And dolphins, if you believe the people who know them best, are the gentlest, gayest, most attractive creatures on the face of the planet.

Our tragedy is that after a few million years in the water we came trooping back to the land, unconsciously carrying an assorted package of marine adaptations along with us. *Homo sapiens* in the twentieth century is, as the old saying has it, neither fish, flesh, nor good red herring; and this lies at the root of more of his troubles than he has yet begun to realize.

Four

Orgasm

At this point we are approaching one of the foggiest terrains in the whole field of behavioural evolution—the problem of female sexual response.

It is now, as Jane Austen remarked about something quite different, 'a fact universally acknowledged' that women can and do experience orgasm. Obviously there must be something very peculiar indeed about this physiological process, or else it would not be necessary to begin with such a statement as that. No one feels it necessary to insist: 'Every reputable biologist nowadays admits that women yawn,' or that 'it can no longer be denied that women as well as men are capable of shivering.'

There have, however, been societies and periods in which the fact of female response was by no means universally acknowledged. Women went to their bridal beds knowing little or nothing of what to expect, and were given vague warnings that the experience awaiting them would be repugnant, but must be endured. Men with the highest medical qualifications pontificated that the very concept of female orgasm was the fantasy of depraved minds, and beyond belief. Havelock Ellis quotes Acton, a leading English authority of the day, who condemned the suggestion that women have sexual feelings as 'a vile aspersion'.

Those days, of course, have gone, and you may have gathered the impression that by this time all doubts and confusions have been cleared up and the whole truth made manifest by twentieth-century scientific enlightenment. However, that is not quite how the case stands at the moment. Suppose we begin by peering into the fog and trying to determine just how thick it is.

To begin with, the investigations of the phenomenon of female orgasm have been confined almost exclusively to the species *homo sapiens*.

In the days of Kinsey, when these investigations were carried out by means of verbal questioning, it is perfectly understandable that this should have been so. It wouldn't be very rewarding to walk up to a cow and ask her in what percentage of servicings she achieved a sexual climax. Even in the era of Masters and Johnson, when the verbal element is largely by-passed in favour of the instrumental monitoring of physical reactions, one can readily imagine that it might be harder to get co-operation from an animal than from a human couple convinced of the scientific importance of what they were doing. It would be pretty hard to convince the cow of this.

However, the ingenuity of research workers could sometimes find a way around these difficulties if they applied themselves to the problem. The interesting fact is that very few of them have yet seriously approached it.

One result of this omission—or quite possibly one cause of it—is a belief widely held that female mammals below the level of human beings never experience orgasm. This assumption is a curious reversal of the nineteenth-century attitude. Men then believed that human love was a spiritual affair and that a woman who enjoyed the physical side of marriage was 'behaving like an animal'. But according to the latest canon, where carnal enjoyment is concerned, woman—though not man—leaves the rest of the animal kingdom far behind. We shall examine this strange theory in more detail later. There is of course not the slightest scrap of evidence for it.

When we move from animals to people, the picture is totally different. A very great deal has been said and written about female sexual response, and a good deal of heat has been and is still being generated by rival experts and conflicting schools of thought. But the area of agreed and established fact still remains astonishingly small. It amounts to little more than my initial statement, the fact universally agreed that the thing actually does take place. Masters and Johnson have exhaustively particularized this statement. They go into minute detail about the physical manifestations that take place in various parts of the body before, during, and after its occurrence; they note, and film, and measure, such phenomena as tumescence, detumescence, flushing, rate of heartbeat, sweating, etc., most of which are paralleled by similar reactions in the male.

They also clear up one previously vexing question. Kinsey had stated that 14 per cent of women claimed that, unlike men, they were capable of multiple orgasm. Rival experts poured scorn on this idea. E. Bergler and W. S. Kroger, for instance, declared: 'One of the most fantastic tales female volunteers told Kinsey (who believed it) was that of multiple orgasm. The 14 per cent belonged obviously to the nymphomaniac type of frigidity where excitement mounts repeatedly without reaching a climax. Kinsey was taken in by the near-misses.' Masters and Johnson confirm the unsurprising fact that the women knew what they were talking about and that Kinsey was right.

So far so good. But outside this small illuminated area, all is chaos. We know it takes place; but in what percentage of women it takes place, how long (in evolutionary terms) it has been taking place, where it takes place, how much it matters if it doesn't take place, precisely what causes it to take place, and why it often fails to take place—all these factors are still being hotly debated.

Men experience something and women experience something, with roughly similar physiological effects, but whether female orgasm is a thing on its own or a pale echo of her mate's ('a pseudo-male response,' as Desmond Morris flatly states), remains unproven. (No one, of course, has been heretical enough to query whether his might be a pale echo of hers.)

Women experience something, but as to whether it is one something or two somethings, controversy furiously rages. One school of thought believes there is a clitoral orgasm and a vaginal one; a subsection of the school claims that the first is infantile and the second a sign of maturity; a vociferous section asserts that only the vaginal one counts as real at all; equally vociferous experts protest that far from being the only one that counts, the vaginal orgasm is a pure myth.

Women experience something, and over the last half century have admitted they experience it, have been encouraged to expect it, have even been told they are entitled to demand it; and men have been lengthily instructed in how to help them to experience it; but still it may fail to arrive, giving rise to debate as to which one of them, if either, should apologize, and whether the blame lies with her frigidity or his lack of expertise or the un-Spockian behaviour years ago of one or both of the mothers-in-law. This argument has not been properly resolved even among the pundits, and you can bet it will be a lot longer before it quiets down in some of the bedrooms.

In case you think I am exaggerating the lack of consensus in these matters, I append a few quotations and opinions.

Concerning lack of response:

Robert D. Knight says: 'Perhaps 75 per cent of all married women derive little or no pleasure from the sexual act.' Kinsey says that only 10 per cent of women are frigid. Marie Robinson hazards over 40 per cent; L. H. Terman 33 per cent; Weiss and English give 50 per cent; Eustace Chesser only 15 per cent. Bergel asserts that frigidity is a problem which 'concerns from 70 to 80 per cent of all women'; while Inge and Sten Hegeler in *The ABC of Love* assert 'there is no such thing as a frigid woman' (p. 55); 'there is no such thing as a frigid woman' (p. 60); 'there is no such thing as a frigid woman' (p. 62).

Concerning the nature of the orgasm:

Freud believed that clitoral orgasm was an expression of immaturity, neuroticism, masculinity, and/or frigidity, and that to become fully mature, normal, and feminine, a woman must learn to progress or 'transfer' from it to an allegedly deeper and more satisfying type of orgasm located in the vagina. Bergler went further and stated that every woman who cannot or does not have a vaginal orgasm is frigid. (One of the factors that makes reading sex litera-ture like playing blind man's bluff is that everyone has his or her own private definition of terms like 'frigid'.)

Kinsey, on the other hand, is a clitoris man, with an equally large following. He states: 'It is difficult ... in the light of our present understanding of the anatomy and physiology of sexual behaviour, to understand what can be meant by a vaginal orgasm. The literature usually implies that the vagina itself should be the centre of sensory stimulation, and this is a physical and physiologic impos-sibility for nearly all females.'

Theodoor H. Van de Velde refuses to make any qualit-ative distinction between the two and compares them to the 'flavour and aroma of two fine kinds of wine or the chromatic glories and subtleties of two quite separate colour schemes'. Other experts regard the two as complementary, while Hastings points out that 'none of the proponents of the transfer theory, Freud included, have stated signs or symptoms by which one may distinguish between these presumably different types of orgasms', and Masters and Johnson report authoritatively that as far as the collateral physical reactions go, it is impossible to distinguish two types.

So is there one or are there two? If one, which is it?

And what gave rise to the myth of the other one? If two, what is the point of having two when men have only one? Which is the major and which the minor? For answers to all these questions, you pays your money and you takes your choice of experts. It's a little like schismatic theologians arguing about the true nature of the Holy Trinity.

There is, again, no agreement as to whether the discovery, or rediscovery, of the fact of female sexual response was a famous victory for women or not. Feminist Eva Figes says hurray: 'When modern woman discovered the orgasm it was (combined with modern birth control) perhaps the biggest single nail in the coffin of male dominance.' But feminist Ann Koedt is inclined to say boo, and has written a tract called 'The Myth of the Vaginal Orgasm', which seems to imply that this particular type is a pure propaganda campaign dreamed up and plugged by lecherous males, in order to persuade women not only to go to bed with them but to make out that they enjoy it no end, and say 'Thank you' nicely afterward.

How much does it matter, anyway? Kinsey: 'Even without orgasm, considerable pleasure may be found in sexual arousal and in the social aspects of a sexual relationship.' Van de Velde: 'It is at the present time impossible to estimate how much unbalance of mind and nerves, and misery in marriage, [is] due to this check and deprivation of complete relaxation in coitus.' E. Havemann: 'Many women never have orgasm yet greatly enjoy every act of sexual intercourse which gives them profound sensual and psychological satisfaction.' G. V. Hamilton: 'Unless the sex act ends in a fully releasing, fully terminative climax in at least 20 per cent of copulations there is likely to be trouble ahead. The least serious consequence is a chronic sense of tense, restless, unsatisfaction.' C. R. Adams: 'If most other factors are favourable, a wife may be happy in marriage even though she is quite unresponsive sexually.* But poor brave Marie Stopes, who combatted so much rabid ignorance and prejudice that she ended by thinking she knew it all, went to the lengths of inducing her husband to sign a document permitting her, as a simple measure of hygiene, to obtain sexual satisfaction from some other source if he ever became incapable of bestowing it.

The truth is that the progress made has been more apparent than real. Freud said at the end of his life: 'If you want to know more about femininity, you must interrogate your own experience, or turn to the poets, or else wait until science can give you more profound and coherent information.' Robert Ardrey, not long ago, threw in the sponge in a mood very like Freud's: 'Our studies of the female in primate societies have not yet reached a definitive level. And until they do, the weary analyst must continue to grope through the trivialities of his patient's past for the sources of problems of timeless dimension; and we must continue to speak only of tendencies, and hazard only guesses, and regard our feminine companions with no more than a newly speculative eye.'

Most fair-minded readers will agree that the status quo as outlined above could reasonably be described as foggy. I can think of only one other subject which in the same period has spawned so many experts all heatedly contradicting one another, and that is economics. Possibly the reason in both cases is that people subconsciously feel that their own personal interest, their power, happiness, dominance, or self-esteem, may be affected by the outcome of the argument. So perhaps the best place to begin a 'newly speculative' approach would be at the furthest possible remove from the human ego, among the animals.

The current assumption concerning subhuman females is that they experience nothing corresponding to the orgasm as we know it. Two main reasons are advanced for believing this, and two explanations are offered as to why the wife of *homo sapiens* went to the trouble of introducing this mechanism for the first time on this planet.

The first reason for believing animals don't have it is that the mechanism in women is so deplorably defective that it must be a recent innovation that hasn't had time to perfect itself by the processes of natural selection. E. Elkan therefore deduces that the capacity for it is still emerging. L. H. Terman, after searching in vain for Freudian hangups among most of his 'sexually inadequate' female patients, is driven to the conclusion that the cause may be biological or genetical rather than psychological. He reaches the same conclusion as Elkan, and Desmond Morris also assumes that even if it is not still emerging, at least it was in homo sapiens that it first emerged.

The second argument is based on the fact that your average female quadruped after copulation strolls away as if nothing had happened, thus plainly indicating that the flavour and aroma and chromatic glories and all the rest of it are a closed book to her.

The alleged reasons for the sudden appearance of the phenomenon in our species are:

(a) Our old friend, the fact that her mate had become a Mighty Hunter and had to cement the pair bond by making sex sexier. She, as the hunter's wife, had therefore to be given a new 'behavioural reward' for being available to him at any time he happened to come back to base. The orgasm is the behavioural reward.

Reason (b) is even more ingenious. It is suggested that when woman became bipedal her fertility was endangered by the new angle of her vagina, which would allow the semen to trickle away and be wasted if she got up and walked away at once. So she had to be temporarily pole-axed by this tremendous and overpowering experience, which would serve to keep her horizontal until the spermatozoa had got where they were going and it was okay for her to get up.

I'm not really convinced by any of this. I agree the mechanism in women is defective, and I'm sure Terman was right to conclude that its defectiveness can very seldom be attributed to seeing something nasty in the woodshed at the age of five. But I don't think it's still emerging. Such things only 'emerge' in a species if they contribute something to the survival of the population. If this mechanism was designed to favour human fertility by overcoming a postulated feminine reluctance to take part in sexual activity, then it is very puzzling to note that when females are young, timid, and fertile, the mechanism is at its weakest; according to Kinsey's figures it doesn't reach its peak of dependability until after fifteen or more years of marriage, when fertility is already declining and coyness must surely be a thing of the past. Moreover, when we think of the size of families in the Victorian era when the mere idea of female climax was considered a vile aspersion, it becomes obvious that if it affects fertility at all the effect must be strictly minimal. In brief, it would only begin to emerge, or continue to emerge, if the women who experienced it were more fertile than the women who didn't. And there is nothing to suggest that this is the case now, or that it ever has been the case.

As for the sangfroid of the quadrupeds, may we not attribute this to the fact that they make far less fuss about everything than we do? If we merely vomit, we are liable to make a fair amount of noise over it, and then sit pale and damp and panting on the edge of the bathtub for five or ten minutes before our metabolism returns to normal; whereas many animals can disgorge almost as silently and neatly as they eat, and immediately 'walk away as if nothing had happened'. We only know something *has* happened because there is as it were an end product. In the same way, everyone appears to believe that male animals experience orgasm because in that case too there is an end product; but if we expected a male chimp to register his sensations by demonstrations of rapture, or exhaustion, or even post-coital tristesse, we should wait a very long time, and perhaps mistakenly conclude that in man too this behavioural reward was only just beginning to evolve.

There are objections also to the idea that a behavioural reward was bestowed on one species alone to make them faithful to their hunting husbands and available to them at all times. Faithfulness, in any case, would hardly be furthered by it: if the reward was so tempting, it could be enjoyed even more frequently by being unfaithful. As for availability, it is true that many of the lower primate females are only receptive at certain fixed and rather brief periods of their sexual cycle, whereas the female of homo sapiens, biologically speaking, is receptive all the time.

But this is by no means due to a sudden jump that was made when her mate began carrying a spear. It has nothing to do with hunting at all. It is merely the culminating point of a gradual extension of the period of receptiveness, which increases quite steadily as we move up the scale from the lower primates to the higher. By the time we reach the higher monkeys and the apes, although the females are still subject to an oestrus cycle, they are usually receptive to sexual advances made at any point of the cycle, and in some cases even during pregnancy. This well-established tendency may have continued to operate as we moved even further up the scale from anthropoid to hominid, but it did not accelerate, and there is no reason to assume that any new factor is involved at that particular stage of evolution.

Finally, there is the theory of female climax as a way of keeping a woman supine, in order to speed the little sperms along their path. I don't go for this one. I doubt if the duration of post-coital lassitude is significantly greater in women than in men. Doubtless most women in practice remain lying down for a while, but then, in our culture there is a strong tendency for sex to take place in bed, most frequently at the end of the day, and in any case at a time and place where interruption is unlikely. There is no incentive for her to leap up and say, 'That was fine, but now I've got to dash.'

I'm willing to bet, though, that however superb a performer her partner might be, if she suddenly smelled burning and realized she'd left the iron on for the last half hour, he would find that orgasm as a guarantee of horizontality is just not on, even today. And among our diurnal ancestors, in the bright sunlight of the savannah, where the process is alleged to have begun, and with the whole troop active all around her, I don't believe she'd have stayed on her back for more than a few seconds.

Suppose, then, we scrap all this supposition and begin at the beginning with a really daring hypothesis. Suppose we get right away from the androcentric concept that sees a world where male animals are created with sexual needs and desires, the attainment of which is attended with sexual pleasure, and female animals are created to serve their needs and facilitate their pleasures and bear their young.

Let us try to imagine a more democratic sort of universe, where nature·or God or evolution or what-have-you looks at the thing a little more impartially, instead of regarding the females as second-class citizens.

The problem was a fairly simple one: how to induce animal A and animal B to get together for purposes of procreation. The answer would seem to be simple, too: Let them enjoy getting together. What conceivable evolutionary purpose would be served by going only halfway to this solution, by making animal A desirous and pleasure-seeking and rewarded with pleasurable sensations, and animal B merely meek and submissive and programmed to put up with it?

Every piece of circumstantial evidence we possess concerning animal behaviour points to the conclusion that the sexual drive is a mutual affair—that both sexes feel a need, both are impelled to satisfy it, and both experience copulation as a consummatory act. Sarel, Eimerl and Irven de Vore point out:

'People do not eat because they need food to survive, nor do they practice sex because copulation is essential for preservation of the species. Nor does a human mother hold and fondle her infant because if deprived of her attentions it would die. We eat, we copulate, and if we are mothers we look after our babies because such activities are pleasurable.'

It is, of course, theoretically possible to regard sex not as a co-operative social bond, but rather as a specialized nonlethal form of predation, and to point out that when a cat eats a mouse, as long as the cat enjoys it, it is not necessary for the mouse to get a kick out of it, too.

Indeed, to judge by their terminology, the predatory parallel haunts men's mind with a curious persistence. In most languages there is some variation on the metaphor that regards a man in pursuit of a female as a wolf, and the girl as something edible—a swell dish, a chick, or a peach.

In the animal world, however, the parallel won't hold up for a moment. The mouse gets eaten against his will; but in no mammal species, with the sole exception of *homo sapiens*, does a female in the wild ever get mated against her will.

Leonard Williams reminds us: 'The male monkey cannot in fact mate with the female without her invitation and willingness to co-operate. In monkey society there is no such thing as rape, prostitution, or even passive consent.'

We can go lower than the primates and find the same thing. Marcel Sire in *The Social Life of Animals* is here describing the brown rat: 'Females, for example, seek males of high social rank; one female is cited who refused to copulate for two months, while a female is normally receptive every five days; the males probably appeared unsatisfactory to her until the day when she met a male of good weight whom she accepted immediately.'

Or lower still, to Drosophlia. Haldane relates an experiment with these fruit flies, where black female flies were confined with males which, owing to a recessive gene, had a yellow body colour instead of black, and a less polished routine of courtship. The males attempted to mate, but the females just didn't fancy them. Ninety-seven per cent of them 'actively resisted, by dodging, kicking, or extruding the ovipositor'. Only 3 per cent laid fertile eggs, and this was not because they had resisted in vain but because, as further genetic experiments proved, they were innately innocent of colour prejudice.

Humanity apart, then, before any sexual approach can be successful, the female must be a willing partner. In many species there are indications that she is even more willing than the male. In an experiment described in 1970 by Dr. Stuart Dimond, both male and female rats were enabled to summon a sexual partner by pressing a lever. At the right stage of their oestrus cycle, females tended to continue pressing the lever even after the males had got fed up with the pastime.

Primates are by no means an exception to this. Robert Ardrey explains: 'In the daily life of the upper primate, the female becomes the sexual aggressor. Someone must initiate the act of love. More often than not, she does.... The appetites are usually hers.'

All I am asking you to deduce from all this is that long before *homo sapiens* arrived on the scene, female mammals were conducting themselves as though sex was to them, as much as to their mates, a desirable and pleasurable experience; that there is no reason to believe, if the males obtained a behavioural reward, the females did not receive an equal behavioural reward. There is no reason to believe that in purely physiological terms the behavioural reward accruing to a man from the sexual act is biologically different in kind from that accruing to a silverback gorilla; and precisely the same is true of their female consorts.

Indeed, there is some reason to believe that the mechanism in both cases is exactly the same. A report from the Kinsey Institute in 1953 records: 'Almost instantly, or within a matter of seconds or a minute or so after the initiation of sexual contact, certain areas of the body may become swollen, enlarged, and stiff with an excess of blood. This is equally true of the human *and lower mammalian species, both males and females.*' (My italics.)

The simplest and most plausible assumption, it seems to me, is that the female sexual response was evolved and perfected many millions of years ago; how far down the mammalian ladder we cannot tell, but certainly in creatures far simpler and more primitive than ourselves. The reason they make so little fuss about it is precisely because it *had* been perfected; it was as simple and effortless as eating, and the reward in terms of pleasure was as automatic.

So the question to which we ought to be addressing ourselves is not: How and why did the human species evolve this frightfully complicated and mysterious female mechanism? It is rather: How on earth did the human species come to lose, mislay, and/or generally louse up such a simple straightforward process?

First, let's try to get a clearer idea of what the process is, and precisely what triggers it off. It would be a help here to forget all about Kinsey and Masters and Johnson and the chromatic glories and all the human hoo-ha, and keep firmly in your mind the picture of, say, a cat or a rhesus monkey. Standing on its four legs.

The answer then, is simple. What triggers it off is a brief but vigorous application of rapid rhythmical friction. That's all it takes.

Some theorists have searched high and low for a parallel

to female orgasm, and the odds-on favourite recently has been the sneeze. The parallel here is presumably that you know it's coming for quite a while before it comes, and you experience a strongish sense of anticlimax if it changes its mind and goes away again. (If it doesn't go away there is no resemblance at all.)

However, if we regard it in its simplest terms as a phenomenon leading to a behavioural pattern designed to relieve it by a brief vigorous application of rapid rhythmical friction, then a closer parallel immediately springs to mind. It isn't a sort of sneeze at all. What we are dealing with is a rather specialized sort of itch. Subjectively, it doesn't feel like an itch, aesthetically we would tend to reject the comparison because its associations are at best undignified; physiologically the parallel is defective because an itch usually arises on the surface of the skin and this does not; but despite all these disadvantages, when we view it in terms of its effect on the animal's behaviour, it is the nearest we can get.

The desired friction is applied, usually from behind, to the interior wall of the vagina. Because of this location, the cat, for instance, cannot apply the friction in its customary fashion with its own hind leg, but must seek the assistance of another cat; and the same applies to the rhesus monkey. It solicits this assistance as simply and naturally as it solicits grooming if there is a flea between its shoulder blades.

The only other point we need note here is that in many primates and other quadrupeds the pressure comes not only from behind, but from above downward, so that it is applied to the *ventral* wall of the vagina. In some primates this is ensured by sexual dimorphism—the male is taller and his legs are longer. In some species, such as the rhesus monkey, where sexual dimorphism is negligible, the downward angle is obtained by the male's standing with both feet on the joints of the female's hind legs so as to gain height.

In the cat it is ensured by the posture of the female—she straddles her hind legs and lowers her spine as near as she can to ground level. And many other animals display a similar behaviour. Desmond Morris observed of the green acouchi: 'When the female is receptive, she stops walking away from the male and adopts the sex-crouch posture. This differs from the submissive crouch in several respects. Firstly, the female's body is lowered but does not touch the ground. Secondly, she flattens her back so that her rear end is correspondingly raised...' etc. It would seem that in mammals, at least, the target is a highly localized one.

If we are right so far, we are now in a strong position to arbitrate one of the classic schisms among the sex experts, i.e., whether the archetypal key to female sexual gratification is centred in the vagina or in the clitoris.

You have only to take one look at the anatomy of a cat or most other quadrupeds to see that as far as most of them are concerned the clitoris has nothing to do with it. There are a few species in which this organ, instead of being tiny and vestigial, is large and prominent. Some animals even have a bone in the clitoris, just as others have a bone in the penis—but even in these species its development appears to have no immediate relevance to the sex act.

No one has definitely explained why it is so highly developed in some species; but Leonard Williams, who is on friendlier terms with one group of these females than anyone else alive today, believes that it may have a closer correlation with urination than with sex. He has closely observed the conduct of his woolly monkeys and suggests that in their case the clitoris has become long and stiff because they are very fastidious about keeping the branches of their treetop homes undefiled, and this endowment enables them to direct a flow of urine away from their own domicile almost as accurately as their brothers can.

Another species famous for the size of its clitoris is the hyena, in which it rivals the length of the penis—it can even be erected in the same manner—so that it is impossible for observers to use this clue to distinguish between the sexes. Yet once again nobody has been able to correlate this with any inordinate sexual appetite on the part of the female. It so happens that in this species the females are more dominant than the males, so that it has been plausibly argued—especially as she also has two balls of fat uncannily resembling a scrotum—that it is an imitative adaptation designed to warn any wandering stranger that she is a tough baby and not lightly to be tangled with.

On the other hand the female woolly monkey is decidedly undominant. The only faint hint of a common factor here is the possible desire for precision for aim in urination, for in the hyena also the urinary tract ends in the clitoris. (In women, of course, it is totally separate.) As a dominant, the female hyena takes a leading part in the usual canid marking of territory. This may have been done in the conventional fashion with urine for a period, though the species has now evolved special scent glands for that purpose; and she still uses urine for spot-marking when she is in oestrus.

If then we eliminate this organ as, in most quadrupeds, sexually irrelevant, we are left with only one possibility, that the centre of gratification lay originally in the vagina. In saying this we run into deep trouble: a concerted protest from the crack troops, Kinsey and Masters and Johnson speaking for once with a single voice, and declaring this to be physiologically impossible. They refer us to one devastating fact; that whereas the clitoris is richly endowed with highly sensitive nerve endings, the inner walls of the vagina for most of their length have no nerve endings at all. The surface there is totally numb. Masters and Johnson with their usual thoroughness verified this point by stroking this inner wall gently and confirmed that the owner of the wall noticed no sensation whatsoever.

This I can well believe. But we have already established that what happened to the cat and the rhesus monkey was not a gentle stroking but a rapid vigorous massage, such as a dog will apply to the back of its ear. Sticking to this analogy for the time being, everyone knows that you can't relieve an itch by stroking it gently. If you've ever had chickenpox and been warned that if you scratch the top off one of those spots you'll be scarred for life, you may have tried stroking it instead, because it's next to impossible to lie there doing nothing; and you'll have discovered that stroking is worse than useless. Whatever mechanism it is that conveys that exquisite sense of relief afforded by a good scratch, it is certainly not triggered off by anything on the surface of the skin, but by something deeper down, which stroking fails to activate.

Suppose then that in this cat, and in this rhesus monkey, and in most other fairly advanced quadrupeds, the trigger of the consummatory sexual experience lies in the muscular tissues just below (i.e., to the ventral side) of the horizontal vagina. Vigorous friction of the lower or ventral wall is what is required, and this is what the male is programmed to apply. Like any other highly evolved piece of behavioural mechanism, it works like a dream every time— *as long as she remains the right way up.*

For our ancestors, this was the ironic proviso. When he turned her over, she was not only frightened and uncomfortable; she was robbed of her behavioural reward. However lustily he exerted himself, he was applying his friction no longer to the ventral surface of the vagina wall but to the dorsal surface; and this had no hinterland of specially sensitized muscular tissue behind it. What was behind was chiefly caudal vertebrae. From the female's point of view the whole exercise was a dead loss. Of course the ape had no idea what had gone wrong. As far as he could make out, all the females of his species had gone cantankerous and completely frigid in a surprisingly short space of time and for no earthly reason.

One fairly inevitable consequence was that oestrus in the aquatic ape began to die out. It was a benign development. There would have been no point in maintaining a periodic peak of intensity in a desire which was not being satisfied. Probably there was a period when those females in which the cycle manifested itself least strongly were the least impossible to live with, or the least likely to be distracted by unassuageable lust from proper care of their infants, and in this way natural selection would ensure that their progeny would thrive, and periodical oestrus ultimately cease to be a part of our genetical endowment.

You might imagine that the capacity to attain a climax would also have died out in the females; but evolution doesn't necessarily work like that. Lamarck thought it did; he thought that any biological endowment which was not being utilized by a particular species would tend to wither away. Scientists today, however, believe that disuse alone is not enough to cause this, and that modification only takes place if some adaptational advantage accrues to the species.

There would have been no particular advantage either to the females or to the population as a whole in eliminating the capacity for orgasm; and so it persisted, and has persisted to this day, even though, in its function as a sexual reward, it may remain dormant, certainly in individuals, and possibly in whole communities, for long periods of time.

It is very doubtful whether it was ever as universally dormant as the pre-Stopesian written records might suggest. If the mechanism is as I have suggested, it is very easy to understand why the thing happens more often in long-standing marriages. In new ones the male quickly attains his own climax and he would not be likely to fire off any response—but after some years, even in Victorian marriages where nothing was expected to happen, the chances are that it did happen, quite frequently.

As his responses became slower and his activity longer sustained, his wife's arousal would become more acute. He could never get the angle quite right—no one will ever be able to get the angle quite right again—but a prolonged friction *parallel* to the ventral surface of the vagina would eventually have the same effect as a short brisk one at an angle to it. If we return to the child with the chickenpox (I am sorry to use so unpoetic an analogy, but poetry didn't come into the business until later)—if the spot itches excruciatingly and he's forbidden to have a good deep scratch and stroking doesn't help, he will eventually discover that if you rub the place with the flat of your hand to and fro rapidly for quite a long time, you can obtain relief even though you're not pressing hard at all.

Heaven only knows what those Victorians thought had happened. Naturally they wouldn't have told anyone; but many a flagging marriage must have received an unexpected shot in the arm just when all the magic had gone out of it. When it had happened once it would happen with increasing frequency, because she would begin to know how to help it happen. This is what is meant by calling vaginal orgasm 'a learned response'.

It is also the explanation of some of the phenomena which gentlemen novelists describe in such loving detail, and so sadly misinterpret. When one of these boys goes into a graphic bedroom scene he is apt to assume that the heroine's frenzied downward pressure on the base of his hero's spine means: 'Don't go—don't ever leave me.' In fact it means: 'I have a subconscious conviction that if you could depress the fulcrum an inch or so lower, this would elevate the tip to where it would do a bit more good.'

When he describes the involuntary arching of the

blonde's spine, he translates it as: 'Oh, God, I'm in extremis, I'm dying of rapture. This is my pseudo-strychnine response, the risus sardonicus, the spinal convulsion.' What it means, at however subterranean a level, is: 'Ah well, if you can't adjust the angle of the piston, I guess it's up to me to adjust the angle of the cylinder.'

It's now time to return to those members of the clitoris school who have been pacing up and down on the sidelines fuming. Because they know their system works. I wouldn't deny it: I would only suggest it is a substitute. The clitoris was a vestigial organ, a homologue of the penis, serving no more useful function than a man's nipples. Like them, it was originally only there because the basic embryological blueprint is ambivalent, and sets out to make a human being pure and simple before it attends to such details as whether the model is to be a boy or a girl; like men's nipples also, it was well supplied with nerve endings because in the alternative model they would have been needed.

However, when the normal sexual mechanism began to malfunction the clitoris was there, and it began to serve a purpose. Desmond Morris states that in a man or woman in a really high state of arousal, orgasm may be induced by merely palpitating the earlobes. It's obvious that if this is so, then stimulating the clitoris would have a similar effect much sooner, because though it was no more originally designed for the purpose than the earlobes, it is much more sensitive and more strategically placed.

The story of evolution is full of such substitutions. For instance, the bones at the angle of a reptile's jaw were economically taken over to form the bones of a mammal's inner ear, since in the mammal they would otherwise have been redundant. If anything in our sexual makeup is still 'emerging', it may be the ease with which this initially irrelevant organ is able to compensate for the malfunctioning of the original quadrupedal behaviour pattern.

The truth is that we are in a state of transition. There is no way back to the quadrupedal Eden. Rear-entry methods are totally unsatisfactory, because although in theory they constitute a return to something nearer the original blueprint, the blueprint has by now been modified past the point of no return (under the pressures of bipedalism the vagina has even developed a self-defeating curve where it bends around the pubic bone), and a woman in this position can make no constructive adjustments at all. The clitoral orgasm is only emergent, and the vaginal one, while possibly the more consummatory, is difficult for some women even in optimum conditions to attain.

This is admittedly all hypothesis. But it is a hypothesis which accommodates more of the facts than most others; and it expains also some of the minor anatomical mysteries, such as why, in those parts of the vagina where the wall itself is sensitive, it is markedly more so on the ventral side.

It was after I had arrived at the preceding hypothesis as the logical answer to some of the more intractable questions, that I first read about Dr. A. H. Kegel, who offered some rather striking confirmation from a quite unexpected source.

Dr. Kegel was a gynecologist working on the problem of urinatory stress incontinence, a disorder which then in severe cases was often corrected by surgery. Kegel discovered that it could be corrected less drastically by a set of exercises designed to strengthen the pubococcygeous muscles that surround and are attached to the walls of the vagina. These exercises not only cured his patients' incontinence, but bestowed on them an unexpected bonus, as reported by Kegel in a letter to the editor of the *Journal of the American Medical Association*. A number of women spontaneously mentioned that their sexual responsiveness and satisfaction had increased—some having an orgasm for the first time in their marriage.

Ruth and Edward Brecher mention this finding in their book *An Analysis of Human Sexual Response*, and comment: 'This led to the conclusion that the physiological basis of vaginal orgasm involved highly specialized nerve endings in the pubococcygeus muscles that are stimulated by penetration of the penis in intercourse, giving rise to a pressure or deep-touch response,' and 'As far as the treatment of frigidity is concerned, Kegel reports that strengthening these muscles through daily exercise is successful in more than six out of ten women.'

In much the same way as *The Naked Ape* took a brief passing glance at Hardy and his aquatic theory, so the Brecher book devotes one page out of 350 to Kegel, as an offbeat theorist commanding minority support. I believe Kegel was speaking the simple truth, and that when this has been more widely understood it will mean that long stretches of the other 349 pages, particularly the accounts of some of the more intricate theorizing of Sigmund Freud, are now of historical interest only.

This, then, was how it came about that the naked ape was faced with a situation which was unique and unnatural—a situation where most of the motivation and most of the reward for sexual activity was confined to one side only—the male side.

Oestrus never returned to the female of homo sapiens. Somewhere in the very bottom layers of his consciousness is a deeply buried conviction that there is something prissy and phoney about the way women carry on, and that if they weren't so damned hypocritical there should be times for every one of them—say one week in four—when she careered round the streets gaily admitting that she was mad for it, soliciting sex from all comers like a young howler monkey, and pursuing her prey until the sun went down and the men were all cowering exhausted in secret male hideouts.

Alas for homo sap., we don't behave like that anymore.

We are not the match for him that we were originally designed to be. We chase after him for love, companionship, excitement, curiosity, security, a home and family, prestige, escape, or the joy of being held in his arms. But there still remains a basic imbalance between the urgency of his lust and ours, so that when it comes to the crunch the prostitute is always on a sellers' market.

I believe this imbalance was not in the original primate blueprint. It's a scar that was left, among several others, to remind us of the 'salutary baptism' that enabled us to survive the Pliocene.

It's not man's fault. God knows it isn't woman's, either. But we may have to wait another couple of million years before the last embers of his subterranean resentment finally cease to smoulder.

Five

Love

'The psychosexual pattern in man,' writes Alex Comfort in his book *Nature and Human Nature*, 'looks very much like the end product of a biological emergency.'

What confronted our ancestors after they had been pushed, by aquatic and bipedal morphological changes, into ventro-ventral sex, was indeed a biological emergency of traumatic dimensions.

It is well known to behaviourists that if you condition a cat to expect that a certain action (e.g., opening a lid) will be rewarded by pleasurable consequences (finding food) and then subject it to conditions where the same action frequently results in unpleasant consequences (a blast of cold air) then you will end up with one hell of a disturbed cat. It will go completely to pieces. It will even take a drink if alcohol is left within reach, and end up on the feline equivalent of Skid Row.

In one respect the plight of the naked apes was even worse than that of the experimental cat, for they had been conditioned to expect sex to be a comforting experience not by a few months of individual conditioning, but by millions of years of evolution. Now the female found that soliciting led not to a well-understood and satisfying response but to an alarming one, barren of reward. And the male found that the warm welcome which had always greeted his attentions was gone. A sexual pass—perhaps a hand on her waist, after the fashion of present-day apes—resulted in alarm and flight; if he pursued her and persisted it had to be in the teeth of kicking, and biting, and screams and tears. And when it was over, so far from the experience having 'cemented a bond' between them, it had only widened the rift, and she was liable to run sobbing back into the sea to get away from him.

(Freud, who thought up explanations for so much of the symbolism that decorated his patients' dreams, never really accounted for the torrents of water that flooded through them whenever their sex lives were growing turbulent. And Lionel Tiger in his *Men in Groups*, after praising the athletic superiority of the male, appends this snide footnote: 'Ironically, the only sport in which females are superior to males is long-distance swimming—a precisely inappropriate skill for a terrestrial mammal.')

In another respect, however, the male hominid was faring better than the neurotic cat, because he was bigger and stronger than the female, and the supine position is a particularly helpless one, so that more often than not he did in fact obtain his customary reward even if he got a blast of cold air to go with it. We must be thankful that he did, or none of us would be here today.

You may be wondering why such an apparently simple and minor biological maladjustment didn't right itself in the course of a few thousand generations. After all, we have up to now been talking quite casually, as evolutionists have a habit of doing, about the most astounding morphological changes in the primate frame, as if endless variations in the shape, size, and arrangement of organs were available through some celestial mail-order catalogue.

Dear Sir,

Am returning fur coat as I have no use for it after all; kindly exchange for 1 pr. earlobes and 14 lbs. subcutaneous fat. The corrugator muscles arrived safely and are satisfactory, but both the brain and the penis are 3 sizes too small for present needs, please replace. I could also use a nose, if you have any in stock.

And oblige, Yrs., N. Ape.

As we know, all these requirements were ultimately met. It seems a little strange that his wife didn't enclose a small petition in the same envelope: 'P.S. Lately my husband has changed his habits and I find the tickly bit of my vagina is now on the wrong side. Have you got a new model. Thanking you in anticipation.'

If she did, it wouldn't have been the first time she got a dusty answer. She was physically a little more complex than her brothers and a lot of her mechanism was badly adjusted to the new mode of living. During pregnancy, for instance, the muscles which supported the weight of her unborn child were all slung from the spine, which was fine for a quadruped, but when she began walking upright everything slipped sideways like an upended line of washing. She would have profited from an entirely new arrangement with the muscles attached to the shoulder bones instead, but though she complained sporadically about backache and prolapse and varicose veins and other feminine disorders, nothing was ever done about it.

In fact, all the evolutionary modifications are governed by two major rules. One rule is that modifications do not take place because they make life easier for individuals, but because they help populations to survive. Think of the agonizing struggles of a giant turtle hauling her exhausted bulk up the beach to dig a hole for her eggs, or a salmon battering itself against the rocks of a cataract, or a crowd of incubating emperor penguins going into a black fast for six howling weeks of an Antarctic blizzard. You can see that in the evolutionary scale of priorities, parental convenience rates very low indeed. As long as her mate's sexual drive remained strong and accurate enough to cause him to rape her, then the fact that she didn't enjoy the process would have no evolutionary impact whatever.

The second rule is that sudden changes in the basic ground plan never take place. The changes are quantitative ones. Our body hair, for instance, never really disappeared: it merely got more and more exiguous. The changes from a reptile's foot to a monkey's hand, a horse's hoof, or a bat's wing were all brought about by gradual quantitative changes in the relative lengths of the bones. So, to expect the deep-touch trigger of the female primate's sexual response to be moved and brought into a quite different spatial relationship with the rest of her organs would be like expecting her mouth to move up on to her forehead. Even the direst biological emergency cannot be cured by such means as this.

So the pair of them were stuck with the unsatisfactory situation. For the first few millennia there would be no danger to the survival of the species. She would still, at that stage, regularly experience oestrus: and probably for a long period, however often she was disappointed, she would continue to solicit, because she knew no other way of responding to the oestral drive. In any case, her mate would receive unmistakable signs of her condition and would respond to them. This is what I mean by saying 'as long as his sexual drive remained strong and accurate'.

The danger was that a time might come when even this ceased to be true. To a certain extent, indeed, it did cease to be true. Here is another pertinent quotation from Alex Comfort:

'Perhaps the oddest biological feature of human psychosexual development is its surprising and un-Darwinian vagueness of aim. Fixity of sexual object, at least to the extent of mating with a potentially fertile female, would seem to be the behavioural minimum to be expected in a system where 'fitness' is a correlate, simply, of total fertile progeny. But the human sexual object is not fixed—under present conditions, at least, it is rather easily displaced.

'Not only is there a large variation, both personal and social, in what female attributes attract the male—there are also gross diversions of sex drive away from its biological object, to inappropriate objects—members of the same sex, articles of clothing, particular rituals or conditions, inanimate objects which, for the unfortunate individual so afflicted, are as compulsively attractive as the normal female is to most men. The majority of these deviations occur in males.

'These deviations, paraphilias or fetishes, are held by Freudians to reflect persisting infantile anxiety.... For the biologist, however, such uncertainty of aim is most odd—it runs wholly counter to what we find in animals, where although maternal response can be imprinted in birds so that they treat a cardboard box as their mother, sexual behaviour is of the lock and key variety; this is what Darwinism would lead one to expect.'

Wherever you don't get what Darwinism leads you to expect, there must be a good explanation somewhere.

We have been slow to find it in this case because, since 'the majority of these deviations occur in males', one might expect that the trouble, whatever it was, arose on the male side. There is also the stultifying tendency to assume that the male has always been the initiator of sexual activity, and the female only a passive instrument or receptacle of his desire.

In fact, among mammals, though the male may be the more active and pursuing partner in the sexual relationship, he is seldom in the last resort the initiator of it. Like any other relationship, it is a mutual affair.

If we take the mother-and-baby relationship, it is the mother who takes the active role, feeding, bathing, diaperchanging, dressing and undressing, sterilizing, laundering, picking up and carrying, while the baby remains passive; but she is never in any serious doubt that it is the baby who has initiated all this feverish activity, and can usually reactivate it at any hour of the day or night by a mewling cry from his cot. She is responding to a specific stimulus. If the stimulus fails to arrive—i.e., if she remains child-

less—she may feel a nurturing impulse stirring in her
regardless; and in that case she may find, like the male
fetishists or paraphiliacs, that it is projecting itself on to
substitute targets—animals, or invalids, or appetitive
full-grown males whose need of mothering was never
adequately slaked.

Sex in the male mammal is similarly a response to a
stimulus; in this case the stimulus comes from the female.
I believe in this case, too, the 'uncertainty of aim' in
homo sapiens set in when the appropriate stimulus failed
to arrive. But this time the failure did not affect individuals only—it was endemic to the whole species. It
was the failure of oestrus.

Here we need to make a clear distinction between two
quite different manifestations of the sexual cycle as it
affects primate females. One is menstruation, a peculiarly
primate phenomenon found only in old world monkeys,
apes, and man. In these animals it occurs regularly after
puberty, and at intervals roughly equivalent to our own—
i.e. monthly—regardless of the size of the animal. It
has no appreciable effect on the behaviour of the female—
though occasionally, in the wild, a female chimpanzee may
notice the blood and attempt to wipe it off, using leaves
as toilet paper—and it has no effect whatsoever on the
behaviour of the male.

As distinct from this, there is oestrus, affecting not
only primates but females through the whole mammalian
order. In some species it occurs annually, or seasonally.
In some species it occurs every few days, sometimes even
at specific hours of the day. In most primates it occurs,
like menstruation, roughly once a month, but falls precisely
between the menstrual periods. Its effect on the behaviour
of both females and males is liable to be dramatic.

In several primate species—e.g., the chimpanzee, and
most of the macaques and baboons—it is accompanied by
changes in a patch of skin in the genital region known

as the sexual skin, which begins to swell after menstruation and by the peak of oestrus is red, shiny, and distended. (This is the condition which sometimes arouses misguided indignation in visitors to the zoos, who wonder why such a painful-looking affliction has not been treated and 'cured'.)

During this period the female herself manifests increased sexual appetite. She seeks the company of males; she solicits. In species such as the savannah baboon, with a strongly marked rank order, her status shoots up sharply. Even the females tend to show more interest in her. In some cases they tend to harass her; in other cases they groom her more frequently. She is allowed to frequent the central area reserved for VIP's—a sort of invisible and mobile simian Royal Enclosure. When her oestrus reaches its peak one of the alpha males (the ruling clique) not only deigns to mate with her but claims a brief monopoly, though a few days earlier as far as he was concerned she was anybody's.

When the oestral phase is over she retires to the lower ranks again, and spends most of her time with the females, ignored by the alphas, and only occasionally teased and bullied by an adolescent male who wants to practice his dominance technique on a not too formidable opponent. She is free to wander about in the sunshine and eat and sleep and forget all about being a sexual object until next time—to her it is strictly a part-time job.

As far as rank goes, she can take it or leave it alone. She never gets as hung up about status as her brothers do, because she knows her turn at the top will come around again, and she won't have to fight for it, either. She'll be carrying a brand-new baby baboon in her arms, and then no female in the whole troop will outrank her.

Obviously, then, in some social primate species, oestrus is one of the most important factors influencing a female's special relationship with every other animal in the troop.

In the great majority of mammal species, social or solitary, it is certainly a prime factor in regulating the interaction of males and females.

As to the precise nature of the signal emitted there has been some uncertainty. In many mammals the signal is clearly olfactory; for instance, if a bitch in heat has crossed his path, your dog doesn't need to set eyes on her to be aware of her existence and her condition. But since the sense of smell is less acute in primates, and because of the gaudy colouring of the sexual skin, some writers have appeared to assume that the signal in monkeys and apes is chiefly visual and behavioural.

Recent studies suggest that it is neither. I quote from J. Herbert of Birmingham, England:

'We investigated this by pairing male monkeys with females whose sexual skin was in the pale condition following removal of the ovaries; as expected, sexual activity was low. Then, a tiny amount of sex hormone was rubbed directly on to the sexual skin, in a dose too small to produce effects if injected. The sexual skin turned bright red, yet there was hardly any change in the male's sexual behaviour. At this point, a small amount of sex hormone was introduced directly into the vagina and almost immediately the male began copulating.... It seems therefore at the moment the sex hormone causes changes in the vagina (perhaps alterations in smell) either distinct from or in addition to alterations in the sexual skin, and that it is these that trigger the male's sexual interest.'

So we have a biological phenomenon affecting the females of a great many mammal species by some kind of hormonal clock, and it results in the emission of a signal, probably olfactory, which arouses sexual appetitiveness in the male. This is the sex initiator, the stimulus to which male sexuality was designed to be a response. And the species *Homo sapiens* has been bereft of it.

It is hardly surprising that some men betray a 'vagueness

of aim', and become homosexual, or get fixated on ladies' underclothes or the smell of rubber or some such irrelevancy. When the biological emergency grew acute, and malignant fate condemned the hominids to live with totally and permanently anoestrous females, the really strange thing is that they retained enough fixity of purpose to keep the race going at all. For many less advanced creatures the end of oestrus would automatically entail the extinction of the species.

Fortunately our ancestors were primates; and in the higher primates copulation has become to an increasing extent a learned activity. Even when the specific stimulus failed him, the naked ape knew what to do. The knowledge of the process had been handed down to him: he remembered. D. H. Lawrence used to speak with the utmost contempt about 'sex in the head'; but in the higher primates, as Harlow's experiments have proved, quite a lot of it is already in the head, and this is truer of *homo sapiens* than any other primate. Without guidance from the upper cortex of the brain, a man's heart, liver, and kidneys would go on carrying out their appropriate functions with their usual smooth efficiency; but his penis would not. Lawrence greatly overestimated its autonomy.

Some writers on the subject have always been reluctant to accept that such a vital mechanism as the oestrus really has deserted us. Marie Stopes was one of these. She was very excited when she read about, and began to write about, the existence of female orgasm, more particularly since she had not at that time experienced it. She was puzzled, as people have always been, at the apparent fortuitousness of it; how it would let twenty women pass and vouchsafe itself to a twenty-first, apparently no healthier nor better loved nor more deserving than the rest—and not even to that one every time.

She tried to carry into her clinics for working-class women not only the idea that parenthood could be planned, but also the concept that marital sex could be a beautiful and uplifting experience for both partners; and she was pained by the number of her clients who gave this proposition the horse laugh.

Whether she read up on the primates, I don't know. She probably did, because once she got her teeth into a subject there was very little she left unread.

Anyway, she reached the conclusion that the whole trouble was that women, even though their bottoms no longer turned bright red, were still, unknown to themselves or anyone else, subject to rhythmical peaks of nubility. The women who found their husbands disgusting, she said, were the ones who were being approached at the wrong time. Ripeness is all, she advised bridegrooms.

She asked the women in her clinics whether they did not find that at certain regular monthly intervals their desire for sexual activity temporarily increased (or their disinclination for it decreased); and she besought them to try hard to remember at what time of the month these feelings occurred and how they related calendarwise to their menstrual periods, and to tell her about it.

There were some discrepancies in the answers to the second question, but she was a woman of compelling personality, and a fair proportion of her clients lacked the strength of character to challenge the premise. 'Y-yes,' they said. 'Perhaps so. Now you come to mention it.' (If she had asked them what time of year instead of what time of month, they would have said: 'Since you put it to me, yes. In the spring, wasn't it? Would that be right, miss?')

Marie eagerly compiled statistics of their answers, and drew graphs, and her books circulated all over the world, and for a time people believed that Mrs. Sapiens was back in business at the old oestral stand. Havelock Ellis, among others, attributed several aspects of female behaviour to periodicity of desire.

I think it was hokum. Later investigations along similar

lines in both Britain and America produced no consistent results. Some women recognized two peaks in a month, others one, while many could detect nothing at all. Among those who said they experienced regular heightening of desire, the times chosen by the majority were not, as in other primates, the times when conception was most likely, but the times when it was least likely. If this was true it was very nice for Catholics practising birth control by the rhythm method; but in biological terms it would be a senseless and genocidal development.

Men have been asked about their cycles, too; and various experts have proclaimed the male cycle to be (a) annual, (b) twice a year, (c) lunar.

Kenneth Walker in *The Physiology of Sex* states that 'in the majority of men it is somewhere in the region of a week'. He's probably right. But I'd say that had less to do with a man's hormones than with the knowledge that he doesn't have to get up early on Sunday.

At the stage we have now reached, the fortunes of the aquatic apes had just about hit rock bottom. They had had one glorious stroke of luck when they escaped from Africa's torrid drought into the marineland of plenty; but from that time on, the problems of going aquatic had been steadily accumulating and had now reached an acute stage of sociosexual tension, lasting long enough to leave psychological scars on both male and females which we have yet to outgrow.

However, this stage did not last forever. Remember, this hominid was a creature adaptable enough to respond to environmental crises by changing from a terrestrial animal to an aquatic one (and later back again)—and from a vegetarian to a carnivore, and from a quadruped to a biped. It was not likely that he would be permanently stumped by the problem of how to copulate with one of his own females without getting spat at every time.

One thing to remember is that while his lovemaking at

that time was necessarily accompanied by some violence, it contained very little hostility. It's not always easy for someone who is being clobbered to realize that the clobbering may be administered without malice. Perhaps the best way to understand his viewpoint would be by means of a feminine parallel. Think of the time when you have had to approach a bawling infant with a spoonful of precious antibiotic fluid. What you say to him is, in essence, what the aquatic ape would have wanted to say to his mate.

'Come on, darling, open wide, you know you have to have it, it's for your own good ... No, it's *not* nasty, I promise you you'd *love* it if you'd only give it a fair try ... Look, *CUT THAT OUT*, you'll make me spill it ... Sweetie? Please? Oh, for God's sake, shut that row ...! It's no good, you know, you may as well give in ...!' And in the end, especially if you are young and impatient, you pinion the baby's arms and resort to forcible measures which leave him square-mouthed and beet-faced and hysterical with rage because he'd like to gurgle all the filthy stuff down his bib again, and he can't.

Our hominid was confronted with an essentially similar problem, except that he was spurred on by a less rational force than prophylactics and his protagonist was only a few pounds lighter than himself. With *teeth*, moreover.

At the end of such an encounter between mother and child, every sensible woman vows she's never going to go through *that* again. There has to be a way of making the baby like it, or at least tolerate it, or at least not notice what's happening until it's too late. She may never find an absolutely foolproof way, but she keeps on trying. And the naked ape would undoubtedly have done the same. It would have been very difficult for him to find a way out if sex were the sole (or even the primary) bond giving cohesion to primate communities.

Fortunately it is by no means the case. For almost all

the higher primates, the enduring bonds have nothing to do with copulation. There is a whole complex network of social relationships, all of them more permanent and durable than sex. There is, first, the cohesiveness that holds the whole troop together, which is analogous to the instinct that holds together swarms of bees and flocks of geese and herds of deer and colonies of rats and schools of whales.

Then there is the bond between mother and child, which in apes may continue well into adolescence. There is the male bond, concerning which Lionel Tiger waxes so eloquent, which welds the males into cohorts. There is the female bond, which he refuses to dignify with the term 'bond', but which causes the females to come together into assemblies of their own. There is the coeval bond, which causes juveniles to cluster together for play and experiment. And there is the specific bond of friendship, which—among apes and monkeys as well as human beings—causes two individuals to seek one another's company consistently, as if they found pleasure in it.

For the most part these bonds tend to reduce fear and hostility and induce mutual trust and relaxation. Also, for the most part, they all have their own signals and gestures and rewards to cement them.

So what the male did in order to reduce fear and induce trust in his mate was to embark on a wholesale borrowing of these signals from other less disturbed relationships and incorporate them into his sexual repertoire. He was saying in effect: 'It's all right, look. I'm on your side. Think of me as your comrade—as your little baby—as your sister—as your parent—as your friend.' Let's examine a few concrete examples.

Beginning with the bond between mother and child, this is clearly of vital importance in all animals where the offspring are helpless at birth, so it is buttressed by a strong framework of behaviour patterns and mutual psychological rewards. The child gets satisfaction from food, warmth, security, and close contact with the mother: the mother enjoys nursing, and the close contact with the child. The child's lips and the mother's nipples are richly endowed with nerve endings sensitive to pleasurable contact and this helps to ensure that the relationship may get off to a good start.

I believe most women enjoy this actual process, though some at various times have been brainwashed into believing either (a) it's lower-class and bovine or (b) it's a sacred duty to the baby, and failure will imperil his health and the mother-child relationship, or (c) it's an obstacle to 'togetherness' because if father wants to take mother out to dine with the boss while junior's still on his ten o'clock feed, this will imperil the marriage. Even one of these myths can make a misery out of what is to an animal pure pleasure; and some women manage to believe all of them at the same time.

You may ask: How do we *know* that an animal gets pleasure from nursing its child and from contact with it? A fair question, since as we have observed one can't cross-examine quadrupeds about their subjective sensations. However, there is one species sensual and obliging enough to pay a vocal tribute to its physical pleasures; and when a cat lies down to feed its kittens you won't only hear some amateur purring from the offspring—the chances are that it will be answered by a more Rolls-Royce effort from their mother.

Cat owners will also know that rubbing a spot beneath her chin usually earns a specially loud and languorous reward, as though that spot were specially designed for being rubbed. It's not easy to see the purpose of this, until next time you're carrying a tray and your cat infuriates you by insistently weaving to and fro in front of your feet, stretching its head up as far as it will go. Of course, if he were still a kitten and you were really his mother (a relationship in which he's got fixated by domestication) this behaviour would not be infuriating but totally adaptive, because he'd be repeatedly massaging with his skull the deliciously sensitive underside of your chin, causing you to be overcome with feelings of maternal devotion and to lie down and be milked.

In monkeys the zones affording maternal reward don't include the chin because the space relations of mother and young are different. But the rewards will certainly not be less for a primate than for a cat, because the demands of young primates are far more prolonged and demanding than the demands of kittens. It is not easy for us to ascertain where they are, because monkeys do not purr; but it is surely likely, for example, that the arboreal mother receives some pleasurable sensation from the constantly shifting pressure and weight of Junior's hands and feet up and down her spine as she moves through the branches. Otherwise the constant hampering burden would become intolerable long before he was ready to fend for himself. For the young, besides suckling, the later rewards of the maternal link include food-sharing.

The chief behaviour bond cementing friendship—and it is a commoner activity between females than between any other pairs—is mutual grooming. It is a useful process—it seems to include deinfestation, and any wounds or lacerations discovered during grooming are carefully picked clean of dirt—but above all it is an enjoyable process. A monkey will solicit grooming by going up to a neighbour and presenting the back of his neck or any other spot he wishes to commend to his attention, just as your dog may solicit patting by shoving his nose under your hand and endeavouring with a couple of brisk jerks to chuck it up on to his head.

The juvenile clusters cement their relationships by playfulness and high spirits and general horsing about, in the mood which in human beings is expressed by laughter.

There are, besides, countless other modes of physical contact among primates expressive of general amity and good will. Embracing is quite a common one, employed with enthusiasm by many species. Woolly monkeys often make a highly emotional ritual out of this; they come together, shielding the eyes with the forearm, puffing the cheeks, making little sobbing noises, and ending in an intimate huddle or embrace. Kissing is another common comradely gesture. Douroucoulis, for instance, kiss one another on the lips at the drop of a hat.

The thing to remember is that among subhuman primates none of these gestures have anything at all to do with sex. The gestures and rituals of copulation are quite distinct and stereotyped. But it seems perfectly clear that the ancestral hominid did his best to incorporate as many of them as he possibly could in his attempt to make copulation once more an amicable and peaceful relationship. The preface to sexual intercourse today is liable to run through the whole gamut of them.

He embraces her and kisses her, as primates do to their comrades. He gives her presents, often of food—chocolates and so on—as primates do to their children. He tries to entertain her and make her laugh, as primates do to their playmates. As the relationship grows more intimate he fondles her breasts and stimulates her nipples, as a primate infant does when suckling. If he's read up on his love-play manuals he may try a little spinal manipulation. He'll pat and caress her and stroke her hair, the nearest he can get to grooming behaviour. He'll hold her tight in a protective hug, as a primate does to its child.

She will undoubtedly derive pleasure from most or all of these activities, and respond to them. And his androcentricity is so unfathomably deep that he is totally convinced she is built the way she is for the sole and simple purpose of making her sexually desirable to him, and sexually accessible to him. Whenever he locates a sensitive spot on her anatomy he dubs it an 'erogenous zone', as though it had evolved for one purpose alone, and that one Eros.

This is about as sensible as calling the cat's nipples and the cat's chin 'erogenous zones', although no self-respecting tomcat ever pays the slightest attention to them, nor ever needs to. In fact there are only two literally, specifically evolved 'erogenous' zones in the human (or, indeed, any other mammal) species. One is the penis, and the other is the vagina. All the rest were designed for other purposes, and have only come to be sexually exploited in mankind because the regular machinery was malfunctioning.

All these new courtship approaches and others are described in *The Naked Ape* and put under the heading 'Making Sex Sexier'. It seems to me they bear all the hallmarks of a very determined drive in the opposite direction—a campaign to make sex less specifically copulative and incorporate into it all the diverse cohesive social elements that had ever made primates behave as if they were fond of one another. I don't think he was making sex sexier, and I'm pretty sure the female hominid didn't think so either. What she thought he was making was love.

We have now taken the reckless step of admitting into the discussion what many scientists tend to treat as the ultimate four-letter word. Masters and Johnson are pretty outspoken, but they know where to draw the line, and you don't often catch them bandying terms like that around. There is a widely held belief that while 'sex' has a nice hard edge to it, 'love' is wishy-washy, and that women tend to talk about love while men tend to talk about sex merely because women are too mealy-mouthed to say what they really mean.

Very few thinkers have seriously explored the hypothesis that perhaps they tend to say love because they tend to *mean* love.

There was probably a time when female orgasm was

altogether absent, because copulation was brief (your average primate is in and out in about eight seconds flat) and the life span was too short to afford much hope of its accidental rediscovery; and oestrus, if not quite gone, was going. The whole behavioural reward for the female at that time lay not in any local assuagement but in the warm diffuse generalized glow of petting and soothing and security and happiness and a desire to please, generated by the frustrated hominid's new tactics, incorporating the elements of parental support, infantile appeal, and comradely good will. This was what was in it for her. This was the new trigger of sexual receptiveness.

This is still a major part of what is in it for her descendants. Today's sapiens male may bone up on sex manuals till the cows come home; he may glean the impression that in this technological age a finger on a nipple should produce results as dependable as a finger on a light switch, and that if it doesn't he must have been issued with a dud model; he may spend years perfecting his technique, but the chances are that sooner or later she will confuse the whole clear-cut issue by some irrelevant question on the lines of: 'But do you really love me, John?'

Last year Dr. Marc Hollender, of the University of Pennsylvania Medical School, published an article in the *Archives of General Psychiatry* concerning a desire he had discovered among women for nonsexual contacts. Certain situations, he said, can lead to a misunderstanding between a couple. When a woman wishes to be cuddled and nothing more, her message may be and often is misunderstood by her husband. She separates her desire to be held from her wish for sexual activity; her husband is much less likely to do so. If sexual activity results she may feel put upon; if it doesn't he may feel she's led him on, only to rebuff him and make a fool of him.

What this type of woman primarily wants to get from bodily contact, says Dr. Hollender, is a sense of security, comfort, contentment, and 'a conviction that she is loved'. Some sections of the press quoted his findings with an air of mild astonishment, as though he had just turned over a stone and revealed a kinky minority of female sexual deviants. One day some pollster will conduct an investigation asking women in general what relative importance they attach to (a) orgasm, and (b) the conviction of being loved, and if they couldn't have both which they would prefer to forgo. (Would you believe a deviant minority of 90 per cent?)

Love as a concomitant of sexual relations is not a recent romantic invention. It had already begun to raise its head, like Venus Anadyomene, out of those Pliocene waves.

It wasn't confined to the female, either. Right from the start, when the hominid put his arms around her and kissed her, it was not only to make her shut that row. As when a mother hugs and kisses a howling baby, it was done out of fondness, too. He didn't like to see her frightened. And the demonstrations of good will and affection, like the snufflings of woolly monkeys, were a mutual affair, and would tend to arouse in him the same warm and grateful feelings that they aroused in her. And while primate sex is a fleeting and comparatively impersonal business, the other primate bonds whose elements were now being incorporated into it were more personal and long-lasting. For both of them the experience was moving nearer to the emotion we now recognize as love.

The only difference between them was that he would get his behavioural reward even if the emotion was absent, while for her it was even truer than it is today, that without this embellishment, or at least some ritual semblance of it, the whole performance seemed singularly pointless and unsatisfying.

By this time they had passed the most traumatic emergency, and were beginning to enter on a new dimension of personal relationship. Perhaps there were even times when, if he had only had the words, he would have compared her to a summer's day. But all this was a very long time ago, before Australopithecus, and the naked ape was only a dumb animal still.

Or was he?

Six

Speech

Our species has been defined at various times as the bipedal ape, the carnivorous ape, the naked ape, the hunting ape, and the tool-making ape; but the one development which more than any other set her on the road to becoming sapiens, the knowledgeable one, was the fact that she became a speaking ape.

This is the great leap forward that set us at an immense distance from all the other primates. In the beginning was the Word. And one of the most baffling question marks hovering over human evolution is how, when and why we acquired the Word.

Such preadaptation as we possessed was only what we held in common with other anthropoids. We were a social species, and to that extent we needed to communicate with one another—but gorillas and chimpanzees are also social species. Because we had passed through an arboreal stage we had acquired a special kind of face, with forward-facing eyes, a muzzle flattened almost out of existence, flexible lips, and flat vertical incisors within easy reach of the tongue—all very useful for the production of labial, dental, and fricative consonants. But there are very few noises we produce which a chimpanzee is prevented by purely physical barriers from making. His vocal cords are very similar to our own and his mouth isn't prohibitively different. Some new factor must surely have intervened to induce us to exploit so brilliantly resources which in all the other anthropoids have remained dormant.

We don't even know when it happened. Many people believe it is a very recent development. J. B. S. Haldane in 1955 reasoned that 'descriptive language' probably only came with the 'technical revolution of the Upper Paleolithic'.

This, however, refers not to the actual origin of language, but to its elaboration into subtler and more productive patterns of speech. The same applies to a similar explanation given by P. Marler in his *Notes on Developments in the Study of Animal Communication*. He attributes this development to greater security. 'Once our own distant ancestors had achieved a more complex society, helping to relieve the individual of his extreme concern with personal survival in a difficult environment, a larger and more varied vocabulary would become increasingly useful. This in turn would permit separation of the complex of information in each distinctive call into a number of sounds, which could then be used as separate items.'

K. P. Oakley has an interesting theory as to why we decided to resort to our vocal cords in the first place. He speculates that 'man's earliest means of communicating ideas was by gestures with the hands', and perhaps 'an increasing preoccupation of the hands with the making and using of tools could have led to the change from manual to oral gesturing as a means of communication'. This is ingenious. But I am not sure that he was ever such a Stakhanovite that when he had something urgent to convey he wouldn't have stopped working to convey it. The earliest communications would in any case have almost certainly been emotive in content—intended to convey anger, warning, threat, appeasement, or sexual desire, and it is unlikely that under the spur of any of these he would have carried on knapping his flints and simply vocalized about them.

The Tarzanists, as usual, appear to assume that there is no real problem here. Man became a hunter, didn't he?—and this explains everything. Desmond Morris: 'Socially the hunting ape had to increase his urge to communicate and to co-operate with his fellows. Facial expressions and vocalizations had to become more complicated. With the new weapons to hand, he had to develop powerful signals that would inhibit attack within the social group.'

Robert Ardrey: 'The hunting life demands division of labour; the male lion flushes the game for the lioness to kill. It is difficult to believe that little Africanus would not have organized his specialists even more finely. Division of labour demands communication between interdependent partners, but the lion was capable only of a roar. Again, I find it difficult to believe that even in the Pliocene days of prehuman experience we did not lay the foundations of human language.'

It is not quite so simple. George A. Bartholomew and J. B. Birdsell, who have a fruitful habit of peering closely at this elliptical type of argument, point out:

'Such group hunting does not necessarily imply a high level of communication, such as speech, or permanence of organization, for it is characteristic of a number of non-primate carnivorous vertebrates—many canids, some fish-eating birds, and killer whales.'

The studies of the Swiss ethologist R. Schenkel establish that wolves, for instance, already use at least twenty-one communicatory signals, of which fifteen probably involve some visual elements, the others being olfactory and tactile. It can't have been very common for the leader of a prehistoric hunting pack to wish to communicate signals more complex than a Master of Fox Hounds conveys today, or than a shepherd conveys to his sheepdogs, but the shepherd can do it all without a verbal vocabulary and often does, because he wants to send his signals over distances where a voice would be lost, but a whistle will still carry; and when a fox hunt is in full cry I doubt whether the verbal units yelled across the fields often need to total more than twenty-one, especially if you discount expletives.

In fact, if we are to judge by the hunting expeditions of primitive African tribes today, it is precisely visual signals which would have been very much at a premium, because success depends so largely on surprise. The successful hunter is the one who never steps on a twig, who always approaches from downwind, and who can keep perfectly still for long periods with his mouth open to reduce the audibility of his breathing. Laurens van der Post's description of hunting with the Bushmen is full of such accounts as the following:

'It was astonishing how sound travelled in that quiet evening air. However silently the polished crocodile or larded hippo took to the creamy water round us, the ripples resounded like flute song among the reeds.... Straight ahead of us rose a gentle yellow island mound with a great glittering lechwe male surrounded by seven does. Our guide motioned the other two makorros back into the reeds ... he signalled to Longaxe to transfer himself to them and took the two of us alone into a jungle of tall sedges at the side. There he put his paddles away, lay down in the prow with his chin over the edge and began to pull us by the shorter reeds foot by foot, slowly towards the lechwe. He did it so well and patiently that a mauve heron came floating low over my head without even looking at us. Once I looked over the side and saw we were going down a line of baby crocodile. I tapped his shoulder to warn him. He grinned endearingly and pointed to the opposite bank.... Finally the guide motioned me to shoot.'

This is the reality of primitive hunting in the jungle. The shot was that of a rifle but it could as easily (and even more silently) have been a poison arrow. There were no 'complicated vocalizations'. The loudest noise in the whole expedition was the outcry of a startled baboon and the Bushman's angry protest (which served no purpose but to relieve his feelings): 'Oh! you thing of evil!

What is the use of us keeping so silent when you cry "beware" so loudly to the world?'

No—a hunting pack may have needed, like the wolves, a multiplicity of signals of communication, but this does not explain why the hominids, any more than the wolves, opted for vocal rather than visual ones, especially when for the hominid, who had to rely more on stealth than speed, vocalization carried such obvious disadvantages.

True, hunting wolves may vocalize in order to terrify their prey, and the hominid may sometimes have wanted to yell to flush his game. But an ape can yell already; and neither a tally-ho nor a war cry would necessarily have brought us any nearer to a noun or a verb.

As Peter Marler observed in 1965, 'The ability to produce new sounds is not unknown in animals, and it seems reasonable to suppose that it could have developed in non-human primates if natural selection had favoured it.... The major problem is thus not to explain how vocal learning started, in terms of neurophysiological mechanisms, but why such learning was first favoured by natural selection.' This still remains the major problem.

Suppose we begin at the beginning. What modes of communication were available to the primates before one of them learned to speak? There was smell, and touch, and sound, and vision. Touch is not too relevant there. Some of the messages communicated mainly by touch have been discussed in the last chapter, and usually, whether we are thinking of a kiss on the lips or a sock on the jaw, they are of a slightly ineffable character which we have not yet learned to replace by phonemes.

Smell is for most of the animal kingdom one of the most basic, most indispensable, and most universal forms of communication. It was one of the earliest to evolve. Even a primitive unicellular organism, such as the slime moulds, can receive chemical impulses from others of its kind, and is able to recognize and aggregate with its own species by detecting the rhythmical pattern in which waves of acrosin are transmitted—a sort of primitive odoral Morse code.

In many mammal species a large proportion of the brain is devoted exclusively to analysing and interpreting smell signals. *Homo sapiens* has a strong tendency to underestimate their subtlety and efficiency because we all of us labour under a major physical handicap: our organs of smell are about as much use to us as a mole's eyes. Nor has any scientist yet been able to supply us with a smelling aid. As G. K. Chesterton's dog sadly commented:

> Even the scent of roses
> Is not what they supposes,
> And goodness only knowses
> The noselessness of Man.

For a dog's sense of smell is not ten times better than his master's, nor a hundred times, nor a thousand times, but much nearer a million times.

As a form of communication it is hard to beat. It can carry over longer distances than we can either see or hear—a Chinese silkworm moth can detect the presence of a female silkworm moth seven miles away. It can convey very accurate information—a minnow can distinguish the water passing over one member of its shoal from that passing over another. It is a powerful medium for controlling behaviour; a toad tadpole is prevented from cannibalizing one of its own kind, not by elaborate appeasement gestures, but because a tadpole's skin when even slightly wounded secretes a substance called by the Germans *Shreckstoff*—fear-substance—which frightens the daylights out of its conspecifics. Above all, it enables an animal to convey messages which can be deciphered in its absence and after quite a considerable time lapse—and we had to wait for the invention of *writing* before we found another way of doing that. The air around us is full of such signals, in the same way as—we have only recently discovered—the 'silent' oceans are loud with the singing of whales and the chattering of shrimps. How and when and why did we lose the power to intercept them?

One reason is that we stayed too long in the trees. Odours are most interesting and varied at ground level, and the earth, especially when damp, is an excellent medium for retaining them. A dog can trot out of its garden gate and perceive at once many of the events of the past few hours—who passed, man, woman, cat, dog, or horse, which way they went, how long ago. He can tell if the dog was friend or stranger, male or female, infant or adult, large or small, sexually receptive or not, belligerent or fearful.

Species that live in trees or in the air aren't nearly so good, for obvious reasons. A flying bird couldn't possibly tell who or what flew through a certain patch of sky ahead of him because chemical particles in the upper air wouldn't stay put long enough. Similarly a sun-baked branch doesn't retain smells nearly as well as the earth does; and even if it did, it would be hard to trace a gibbon, for instance, across its vast interaboreal leaps by means of a scent trail.

So the birds and the primates traded in this portion of their heritage for better sight. The olfactory lobes in the brains grew smaller and their vision became relatively more important to them. Colour vision in particular seems to bear some correlation to life above ground level. Most birds, and many primates including ourselves, have retained the capacity to distinguish colours, while most of the other mammals have lost it. (It is all nonsense about the bull seeing red.) True there are a lot of colour-blind primates, but this is likely to be because they are the nocturnal ones, and colours don't show up well by moonlight; and there are a few colour-visioned non-primates, but this is likely to be because—like the squirrel—they are just as arboreal as monkeys. This is why birds and primates sometimes deck themselves in vivid reds and blues and yellows to impress their colour-visioned mates, whereas earthbound quadrupeds never wear primary colours.

And, as Haldane pointed out, this is also why, although birds are zoologically so far removed from us, we feel we understand their social behaviour and patterns of courtship, because they are based like ours on auditory and visual signals which we can perceive, while that of mammals is conducted through odour signals, in a language to which we are largely deaf and blind.

But we have to be very careful not to carry this reasoning too far. An ape is by no means such a slouch as we are in the olfactory field, and neither was our ancestor when she left the trees. Some South American monkeys still scent-mark territories as assiduously as beavers do. And if the Tarzanists had been right and our first move had been down to the plains, we should have expected our scent organs to become more important to us, not less. Even birds who return to ground level—such as ducks— reactivate their sense of smell into perfectly serviceable working order.

The primates retained their sense of smell not primarily as a means of perceiving their environment—their eyes were more efficient for that—but as a way of communicating, among other things, their state of mind.

Try to imagine you are a potto or a slender loris proceeding along a branch and you meet another loris head on. You may wish to express annoyance in order to make him get out of your way. You may wish to express apology and appeasement, because he's bigger than you are. If you're female you may wish to express how glad you are to see him. But a loris's face is not very expressive, and you don't have a large selection of noises. You can't do much gesturing because you need your four hands for hanging on to a branch. You can't alter your spatial relationship very much—for instance by giving him a wide berth, or rolling over in submission—because the branch is narrow. Luckily nothing of all this is called for. If the smell of anger comes out of him you will know it; if the smell of fear comes out of you he will know it; and oestrus is an olfactory signal always.

Almost certainly this system was still of greater importance to our Pliocene ancestors. Indeed, there are traces of it in us even today. For example, a woman can readily detect the odour of exaltolide, a substance with a chemical constitution similar to that of civet (and presumably conveying the essence of ♂). She cannot smell it when she is a child, and she cannot smell it after the menopause; but a man can never smell it at all unless you give him an injection of oestrogen.

What about sound? Biologists looking for the birth of language have worked hard on the vocal signals of the higher primates. They are many and varied—howls and growls and snuffles and sobs and roars and screams and snickers and chuckles and lip-smacking and teeth-chattering and whining. Many of them are more or less specific not only to a species but to a definite emotional context— the roar for rage, the lip-smacking for sexual overtures, the snuffles for friendship. After long study of any particular species an observer can frequently make finer distinctions and attach meaning such as: 'Welcome—I am delighted to see you.' 'Please give me your attention.' 'Stop fighting—I give in!' 'Help! Mother!' 'Look out— danger!' and so on; or he may note that a particular sound at a particular pitch and frequency is uttered only when a baby monkey has fallen out of a tree. Even at this level the language is very much more flexible and meaningful than, for instance, birdsong. But the anthropoids have a medium of communication far more flexible and meaningful still—and that is the visual one.

They have, for one thing, their faces. Leonard Williams:

'The face of the woolly monkey is so expressive that one monkey can understand and interpret the intentions and moods of another simply by looking at the other monkey's face.... They are artists at calculated mime. When they are stared at accusingly they will screw up their faces and squint with embarrassment and annoyance, with their chins lifted high, their eyes almost closed, and their teeth bared in an apologetic but disdainful grin.' And the faces of the apes are more expressive still.

Also they have their bodies, and every inch of them is eloquent—the tilt of the head, the carriage of the tail, the stiffness of the spine, the angle of their approach, the speed of the movements, the sweep of their arms, the rise and fall of the hair of their head, the tightening of the scalp. These are not stereo-typical ritualized gestures of 'aggression' or 'appeasement'—every single element in them is subject to infinite gradations and nuances and combinations, all significant of the animal's state of mind.

They have their eyes—it would be possible to write a whole chapter on nothing else but the ocular social signals of primates, from the total attention afforded to the movements of an alpha male to the complete 'cutoff' of looking ostentatiously the other way.

It would be easier still to write another whole chapter on the intricate system of spatial relationships they maintain towards one another. It is not merely that each may preserve around himself an area of what Jane Goodall called 'personal space', varying in extent with his status, and not to be infringed save for the purposes of grooming or sex. It is much subtler than that. An observer who knows the personalities and rank order of three monkeys in a cage would be able to predict with fair accuracy the shape and size of the triangle which would connect the positions of the three after a banana had been offered at a point X on the bars of the cage.

Now, all this adds up to a system of intraspecies communication of an extremely high order—a sensitive and flexible combination of olfactory, vocal, and visual signals, infinitely more subtle and adaptable than that of the wolf. If and when the ancestral ape became a predator, it would surely have been equal to any demands likely to be made on it.

Nevertheless, suppose we assume that the exigencies of the hunting life demanded that the system be brought to an even greater pitch of refinement. Let us temporarily accept that it might have been useful to him to have a signal that signified 'eland'.

Which of these three channels of communication would have been utilized by a hunting anthropoid to convey the signal 'eland'? It obviously wouldn't have been an olfactory one. Odoural signals are not transmitted by design—they are involuntary physiological responses to hormonal or emotional stimuli.

It would not have been a vocal one, either. The reason is precisely the same. In most mammals, including non-human primates, vocal signals are quite as involuntary as odoural ones.

You may train your dog to the point where, in response to a quiet neutral-voiced command 'Sit' or 'Come' or 'Stay' or 'Lie down', he will sit, come, stay, or lie. However patient you are, or however intelligent your dog, you will not train him to the point where, in response to a quiet neutral-voiced command 'Bark' or 'Whine' or 'Growl', he will distinguish between these commands and carry them out. It is not within his power to command these things. You may induce him to bark by putting on a show of excitement yourself, or taking out the lead and moving to the door, but this only means you induced in him an emotional state of which barking is an involuntary concomitant.

With really intensive conditioning, even these 'involuntary' reactions can be induced; for example dogs have been conditioned to increase salivation, or even to inhibit salivation, in response to a stimulus or to attain a desired reward. In the same way, with assiduous and expert training, a dog can be taught to 'speak' on command, i.e., to utter a curiously tentative kind of bark. But teaching this is difficult, in the same way as teaching a deaf child to speak is difficult; the normal dog, like the deaf child, behaves as if it were not getting 'feedback'—an immediate automatic awareness that it has done the thing required of it, and an ability to modify and refine the vocal response as a sheepdog, for example, can modify and refine the physical responses of crouching and stalking and out-flanking.

While the noises made by primates show greater variety than a dog's, they are no less involuntary. All the experimenters who have put so much hard work into attempts to teach apes to speak have been as ungainfully employed as if they had been Martians trying to train men to dilate their pupils, or to blush, or to have an erection, in response to a word of command. However they stepped up the system of bribes and punishments, they would have an uphill task.

The most powerful conditioning agent yet discovered by biological researchers is a fine wire capable of applying electrical stimulation directly to the 'pleasure centres' of an animal's brain. A rat which is enabled to afford itself this stimulus by pressing a lever will continue pressing it unceasingly until it becomes exhausted. Animals can be taught to perform complicated manoeuvres with the incentive of obtaining this supremely compelling reward. But a monkey is simply unable to learn to attain it by emitting a voluntary sound. It would not have to pronounce a word, or any specific syllable; nothing clever at all. It would only have to make a vocal noise. It just can't get the hang of it.

It is true that two American psychologists, K. J. and Caroline Hayes, did teach their chimpanzee Vicki to speak.

Vicki achieved four words, not very clearly spoken, and it cost the three of them six years' sweated labour to achieve this result. In the last resort it is possible for us, and clearly for primates, too, to obtain volitional control of a normally involuntary process; Vicki, given sufficient incentive, learned to emit noises at will, just as a yogi can learn to slow down his heartbeat or reduce his blood pressure at will.

There are, of course, much easier ways of getting chimps to communicate. Allen and Beatrice Gardner, of Reno, taught their chimp Washoe sign language, and within two years she had a vocabulary of thirty-four signs. Her voice could communicate only emotions, but her hands and arms could name nouns and verbs and specify her wishes comparably with a deaf and dumb child of the same age. She even coined her own phrases. The alarm clock which signalled her dinnertime she named 'listen-eat', and she called the refrigerator 'open-food-drink'.

It seems to me quite certain that if an arboreal primate had come from the trees to the savannah and needed badly enough a signal for 'eland' he would have used the primates' long suit, the visual signal. He would have been like Washoe; he would have been like an English hunter with a Chinook guide who wants to find a moose, and spreads high his palmate hands like palmate antlers. He would have *mimed* an eland; as the ceremonial dances of primitive tribes still mime the movements of their prey and their predators.

In practice, I do not believe he would even have felt the need for such a signal. He was an animal. The idea 'eland' would only come into his head if he smelled an eland on the breeze, or heard it stepping through the bush, or saw its slot on the earth or its head against the sky. And in this situation, if he wanted to communicate the idea of it to a companion, all he would need would be a sound or movement to command attention: the equivalent of 'Psst!' His companion would note the cock of his head, the elevation of his nostrils, the direction of his gaze, and follow his example, and anything the first animal perceived would be perceived by the second.

As long as these conditions prevailed he would never feel the need of speech. Certainly he would never have embarked on Vicki's laborious path of teaching his howls and grunts to bend to his will, unless some powerful environmental change had caused his other communicatory channels—the scent glands, the facial expressions, the bodily posture, the gestures, the eye movements, the spatial relationships—*all, simultaneously, to malfunction,* so that only the unlikeliest and least tractable tool, his voice, was left to him to rely on.

This was what happened to him. And it happened long before he became a hunter. When he took to living in the sea olfactory communication became virtually non-operative. This may have been another factor contributing to the end of oestrus; and it would have applied also to the signalling of any other emotional reactions.

Any chemical particles emitted by an aquatic primate would have been very quickly washed off into the water. Now, fish can easily detect odours dissolved in water; insects and animals can detect odours suspended in air; but a lung-breathing primate smells by inhaling air, and if he begins inhaling water he's in trouble.

Far more disconcerting than this, however, was the distortion of the visual signalling system. When you are swimming you can't draw yourself up stiff-legged; you can't make a swift controlled forward dash for two yards and then stop dead; you can't maintain unflinching eye contact with your antagonist with the odd wave sloshing over your head or an undertow hauling you backwards; you can't appease him by presenting or dominate him by mounting; you can't loom over him at your full bipedal height; you can't humiliate him with a 'cutoff' if you can't be sure he'll be watching you or won't attribute your head turn to a new style of crawl stroke; you can't observe or expect him to observe the proscribed amount of personal space of the proper deployment of equilibrial spatial relationship in a medium which can toss you both around like a couple of corks; and some of your most classic facial expressions like the open-mouthed threat face are apt to end in a gurgle and splutter if you hold them too long.

The sea was safe, and cool, and teeming with food; but it was playing hell with social relationships and the dominance structure. Gradually, over many generations, it became borne in on the hominid, as on Vicki, that the only time he got his reward while in the water was not when he bristled or scowled or swung his arms, but when a noise came out of his throat, a phenomenon in himself which he'd never paid much conscious attention to, although he'd recognized the noises when they came out of his companions. The reward in his case was the attention of others of his troop. Like Vicki, he had the incentive to work hard on bringing this physiological function under his conscious control; like Vicki, he found even his best efforts didn't amount to much. But they gave him a bit of an edge over his companions; among primates it is the males who best command attention that tend to become dominant, and it is the dominant males who produce most of the offspring. The individuals who were good at producing noises at will were the most genetically successful, so that in the end everybody could do it.

Later, when he did in fact become a hunter and wanted a signal for 'eland' he chose a vocal one; but that was only because ten million years of aquatic evolution had transformed the vocal channel of communication. From being one of the least likely to expand and diversify, it had become the most likely. To no land mammal has this ever happened.

Sea mammals? Yes, of course. They encountered the

same problem and they came up with the same answer. For a large number of them vocalizing has become a consciously controlled activity. It doesn't take a couple of trained psychologists six years' slog to train a dolphin to do his party piece. A complete tyro can embark on a training course with a supply of fish and a modicum of patience and after a few months his dolphin will obligingly warble into a microphone at the drop of a herring, and have the spectators rolling in the aisles.

For dolphins the vocal-acoustic channel of communication has become the prime one, to a much greater extent even than in men. The olfactory lobes of the brain, much diminished in man, have in the dolphin shrunk almost out of existence; and they use sound not only, as we do, for intraspecies communication. By means of a sonar system they use it for exploring their inanimate environment—that is, as a sort of auxiliary eye.

(It is probably no accident that the bat is the only land mammal who does this. A combination of flying, and hanging inertly while at rest, would go far to invalidate both olfactory and positional signals, and would encourage vocal control, a necessary preadaptation to the sonar system.)

Dolphins have most of the prerequisites of speech. Firstly, they can vocalize at will. Secondly, the noises they emit are many and varied, and carry over long distances. A dolphin's vocal repertoire includes clicks, quacks, wails, whistles, and singing noises both sonic and supersonic. The humpback whale adjusts the pitch of his song to the distance he requires it to travel: high-pitched for communication at close quarters, and low for long-distance transmission.

Thirdly, they have 'feedback'; they listen to their own voices. They will learn to make airborne sounds on request, and to recognize the range of human hearing; they will limit their utterances to a pitch audible to us.

Fourthly, like children, they have the instinct and the

ability to mimic sounds that they hear.

J. C. Lilley reports of the dolphin Elvar: 'After he had heard only human voices for several weeks, his vocalization began to be less "dolphinese" and to break up into more humanoid, word-like, explosive bursts of Donald Duckish quacking.' Dolphins are said to imitate human laughter, especially a woman's, on hearing it, and to utter phraselike sounds that wishfully thinking auditors receive as meaningful syllables.

Finally they have what may or may not be a fifth prerequisite of language: they have a brain of a size comparable to our own, and the size of our own has often been attributed to the development of human speech. We are naturally reluctant to use this reasoning about the dolphin's brain and we tend to explain it away rather hastily as a by-product of echo location: the neural mechanisms of this very complex sonar perceptual system are bound to take up a great deal of room and that is why the dolphin needs such a large skull. (Or is it? And if that is the reason, why doesn't the bat have a bulging cranium, too?)

Nevertheless, all this put together doesn't prove that a dolphin can speak, even to another dolphin. It may mean only that the dolphin is at the same stage as a human infant, whose noises are voluntary, eloquent, imitative, and often—to his mother, at least—perfectly intelligible; yet, as the word 'infant' literally implies, they are still noises and not speech.

There is one more step to go before we can call it speech. A reporter for a Sunday paper colour supplement was recently told by a dolphin trainer in a Florida seaquarium that out of a dolphin's varied utterances a hundred different sounds had been identified, and it had been established that 'twenty-seven of them concerned feeding'. If this is true, it may still mean very little. If they are all variations of such themes as 'hungry' or 'yum-yum' or 'ugh', or 'more', it would only denote that a dolphin can express subtler gradations of the emotions of discomfort, pleasure, and desire than a chimpanzee can express. But if we ever discover that out of the twenty-seven dolphin food noises one means 'herring', then we will know for certain that we are not the only linguists on this planet.

Only one point in all this is strictly relevant to the theme of this book: apart from ourselves there is only one group in the whole of the animal kingdom in which the possibility of such a discovery is being seriously canvassed, even by the wildest of optimists. It is a group of mammals of social habit which moved back from the land to the sea. It could, of course, be just one more of a long long string of coincidences. I don't believe it is. I believe the same forces that brought the dolphins to the very threshold of speech were the forces that brought the hominid to the threshold of it, too, and left all the rest of the primates an immeasurable distance behind.

What pushed us over the threshold, when every other species in water as well as on land remained the other side? Partly the fact that we were by far the most complex and advanced species that ever went aquatic. Our social organization was very highly developed—among aquatic species only the dolphins and killer whales come anywhere near us in this respect—our system of signalling was immensely subtle and expressive, and when those signals ceased to work it was correspondingly imperative to augment them. We used our mouths to command attention, to exercise domination, and to regulate relationships, where previously we had used our faces and our bodies.

But we still have to explain the first noun, the meaningful signal that the land-dwelling ape had never needed to evolve.

It is just possible that in our highly specialized littoral environment this need would now arise. Suppose the ape

was diving around in shallow water and he saw a dugong.
They were very common at that time, because until the
naked ape happened along they had no enemies. It was a
large eight-foot specimen and he needed help with it so he
yelled out 'Hey' to his brother sitting on a rock. His
brother looked at him—he shouted again and beckoned.
But the brother didn't feel like coming in for a swim, and
there was no way of getting him to understand what the
excitement was about. He could see nothing but the waving
hominid and the sunlight sparkling on the surface of the
water. It was not enough to say 'Psst' and point. It was no
longer true, as it would have been true on the savannah,
that what the first animal could perceive the second animal could also perceive.

The one in the water was on fire with the need to
communicate his exclusive bit of information, and the
newly flexible vocal channel was the one to use. We
don't know what noise came out of him—perhaps a yum-
yum noise meaning 'this is food'; perhaps a deep rounded
noise meaning 'this is huge'; or a squishy noise to imitate
the squelching flubbery sound made by the shapeless
creature when it's hauled out on to wet seaweed. For
convenience's sake, let us arbitrarily assume he said 'fish'.
Of course, it was useless. All words are useless until at
least two people know what they mean.

But he had been goaded by the need of a noun as no
animal had ever been goaded before. His failure
smouldered in him, and the next time a dugong got washed
up he grabbed his brother by the hair and pushed his
head down near the smooth wet hide and spat out repeatedly 'Fish! Fish!' and kicked the sirenian and smacked
the brother's head to teach him not to be so stupid next
time. Given a few million years of this type of incident,
sooner or later somebody's brother will get the message;
and the pair will make a very good and adaptable team,
and their tribe will learn from them, and those who are

quickest to understand words will be the likeliest to survive and propagate their kind.

Still, he only said 'fish' when a fish was there; it was not yet a free-floating common noun, but attached itself always to a concrete individual specimen currently present in the vicinity. There came a night when he had retired to his cave. His infant daughter was happy; she was spontaneously vocalizing at random, as a happy young dolphin will do, and practising noises she'd heard on the beach that day. 'Fish,' she said. Her parents laughed because it came out so clearly, and because they laughed she said it again. There was a brief numinous silence because a new mysterious thing had happened. There was no fish in the cave; no fish or bone of the fish, no sight or smell or sound of the fish; but a fish had come into their minds, a common noun holding the essence of all its kind; not the end product of sensory signals from their physical environment, but for the first time generated out of nothing by hominid mental and vocal interaction.

The baby prattled on until her father growled and moved away to sleep, and her mother pushed a nipple into her mouth to quiet her. She wasn't hungry enough to suck, but she liked holding it in her mouth. She went on crooning to herself, sometimes closing her lips around it and sometimes letting it go; thus coining the classic disyllable which has given its name to the whole mighty biological order which had produced her.

'Ma-ma,' said the baby. 'Mama.'

Seven

The U-Turn

Finally, the rain came. Over a long period which archaeologists have never quite pinpointed, the Pliocene merged into the Pleistocene.

The deserts grew green, and the game multiplied, and the shrunken rivers that had trickled through stony beds to the hominid's littoral footholds grew wide and smooth again. The water in them was sweet to drink; bushes and trees sprang up along their banks. There was fruit on the trees and fish in the rivers. The naked apes found the rivers very pleasant and followed them up, farther and father into the interior.

They were no longer afraid of the interior. Life wasn't so fierce now that food and water were plentiful. Besides, they were probably a bit larger than when they left (creatures that take to the sea almost invariably grow bigger) and they were more confident. They had reason to be. They had learned to eat practically anything; they had tools and weapons; and they were masters of two elements, which was a great advantage.

Geographical barriers insuperable to gorillas and chimpanzees were nothing to them now. If they felt the urge to wander they could walk across great plains and swim across the widest rivers. Periodically during the Pleistocene the Great Ice Ages of the northern hemisphere locked up vast quantities of the earth's surface water into icecaps and glaciers, so that the sea level dropped and new land bridges appeared; the hominids crossed those, too, and penetrated into Europe and Asia.

Some of these migrations were so extensive that mid-

Pleistocene remains of near-human species such as homo erectus have been discovered as far afield as Choukoutien in China, and in Java; and this Asian species differs in several respects from most of the African Australopithecines.

I have said they crossed the newly exposed land bridges, because it is commonly assumed that the migrations northward and eastward took place by land and could not have begun before the first Pleistocene glaciation. This is perfectly possible, and of the Far Eastern remains discovered to date none is so ancient as to disprove this assumption. On the Hardy hypothesis, however, this dispersal may have begun earlier still.

At the highest point of their period of aquatic adaptation the ancestral hominids, though never as fully marine as the dolphins or sirenians, would probably have been capable of crossing wide stretches of water under their own steam; and without postulating that at such an early stage of their evolution they became boat-builders, it is highly possible that they would have been aware of some of the uses of a floating log. If we can believe that homo sapiens crossed the Atlantic and Pacific on rafts of reeds and balsa wood, it is possible that his less human but more aquatic ancestors crossed the narrower gap from Africa to the Eurasian land mass while it was still separate, and that it was these colonists who evolved into homo erectus. If Asian remains dating back to the beginning of the Pleistocene are ever discovered, we may be driven to assume that it must have been so.

But wherever the hominid beached, he now had new problems confronting him. Land mammals, he found, were faster and fiercer than the mammals of the sea; but he had to face them, because now that the long heat wave was over, he often found the winds blowing chill on his naked skin, and he needed their furs as well as their meat.

Fortunately, he was equal to the challenge. One great

environment change—from land to sea—had given a great shake-up to his whole biology, introducing new problems, new tensions, remoulding his body, reshaping his methods of apprehending his environment and his fellows, keeping his mind alive and plastic. A second such change in an evolutionarily short time—back to the land—repeated the shake-up before he had time to become overspecialized and settle into the kind of evolutionary rut that engulfs a species too perfectly adapted for a single niche.

You may have observed the pronoun has changed to 'he'. This is because we are approaching the period when male behaviour begins to change faster than female behaviour. The hominid begins to think of specializing in hunting: Tarzan, at long last, is waiting in the wings.

Before he comes on, let us devote one more chapter to looking back at the ocean and its influence.

The theme of this book has been to suggest that a surprisingly large number of the differences between ourselves and our nearest zoological relatives may be explained by postulating that after initially evolving as a land-dwelling mammal, we returned to the sea and became aquatic. This single move is not much in itself to ask any zoologist to swallow, because it has happened over and over again in the history of the world. But the theory further postulates that after going quite a way towards becoming an aquatic creature, one species performed as it were a spectacular evolutionary U-turn, and came back out of the water again and returned to the land permanently.

This has not happened over and over again. The question is whether it ever happened at all to any species other than homo sapiens. After all, there have been animals wandering the world for a very long time now, some of them of more ancient lineage than our own. And there were great climatic upheavals, droughts, and Ice Ages, long before the Pliocene and the Pleistocene. There must have been species more primitive than primates who found the land growing inhospitable and took the water escape route —was there ever one that returned to its old haunts when the earth began to smile again? It would be nice if there was. It's always easier to believe in a thing that has happened twice.

When we look around the world's land animals searching for a fellow revenant, we can narrow the field at once by discarding anything with a coat of lustrous fur. The fur would have been the first thing to go.

I suspect that all the hairless mammals (except freak mutants like the Mexican dog) have gone through a soggy time at one point or another. Some of them, like the hippopotamus, are still happily wallowing in it.

We might consider, for instance, the pig. He's not entirely hairless (neither are we), but you still couldn't call him furry. Even those of his relatives who do grow fur don't do quite a first-class job of it—it's coarse and uneven, the fur of an animal who once lost the knack of growing the stuff, and now is trying hard to remember. The domestic pig doesn't even try very hard. He has, like so many aquatics and ex-aquatics, a layer of subcutaneous fat to keep him warm, and those who should know were wont to report that his flesh tastes very much like our own, especially the rind. It is not too surprising that cannibals are said to have called man-meat 'long pig', for our skin, like the pig's, has the same sparse hair, the same thick epidermis, the same high content of elastic tissue, and Weinstein in 1966 established that it contains similar proteins.

The pig lives on land now, but with occasional signs of maladaptation. For instance, it was observed in America that decent civilized pigs who always use the proper dunging passage for their natural functions when the temperature is under 20 degrees will in really hot weather urinate and defecate in their living quarters and roll over in the stuff, not because they like to be dirty (pigs don't like dirt, and will learn to operate a press-button flush if you provide them with one) but because their skins were specialized for wallowing in, and above a certain temperature the craving to moisten them becomes intolerable.

However, the pig can't count as a revenant in our sense. There is no sign that he ever went to sea. He can, like most mammals, swim a little way if his life depends on it, and the legend that a pig trying to keep afloat will cut his throat with his trotters is nothing but an old wives' tale. But he bears all the hallmarks of a swamp-dweller: the inadequate-looking legs (because in the mud his body would be comparatively weightless and he wouldn't need speed), and the flat damp disc-like nose, like the manatee's, for rooting out vegetable food growing under the mud (it works just as well for truffles growing under loose earth).

Swamp-dwellers have occurred and recurred in large numbers throughout the earth's history. There was an awful lot of mud around when the world was young. Many species of dinosaur lived all their lives in that sort of environment, and later, in the Pleistocene, animals specialized themselves for it in fantastic ways, as, for instance, Hippopotamus gorgops, who wore his eyes on stalks, like periscopes, to keep them out of the mud.

All this has little relevance to sapiens, for this was never the life he led. To sink into a bog for a few million years and then trot out again cannot be compared to our own elegant U-turn. But I believe that one of the pachyderms went further, experienced like ourselves a marine baptism—millions of years before it happened to us—and returned to the land. I offer you the elephant.

We know that most elephants are virtually hairless; we know too that their babies are born tufty enough to confirm that they had fur once upon a time and there must have been some reason why they lost it.

We know that one of the elephant's nearest kin went into the sea and stayed there permanently, for the elephant's closest cousin is the sea cow. We know that early species of elephant developed weird and pointless-looking dental arrangements quite useless to land-dwellers. There were, for instance, the shovel-tuskers, Ambelodon, and Platybelodon grangeri. Now shovels and spoons as a natural endownment are invaluable to water-feeders, like ducks and spoonbills and platypuses, but why on earth would a land-dwelling animal want to scoop up a shovelful of earth? Another primitive elephant had tusks that pointed down, like a walrus's. We don't know what he used them for, but the walrus find them very handy for hauling out of the water on to rocks and icebergs.

However, it is today's elephant that we must study to determine whether the sea has left its mark on him. Look at the size of him. He came from an ancestor two feet high. Why did he achieve a bulk so ponderous and needing such vast quantities of sustenance? (All the land animals of greatest tonnage have lost their fur.) A wild elephant often spends eighteen hours a day just feeding himself. Surely he did his growing, like the whale, in an element where weight is no hindrance and size a positive advantage for heat conservation.

Look at that trunk. No one has adequately explained that trunk. They say that when he grew taller his mouth became farther from the ground, and his tusked head was so heavy that he couldn't afford to grow a long neck, so the trunk was to reach down and pick up food.

But this is no sort of explanation. If he was a grazer, and the grass was as short as all that, he could never have picked enough blades of it in his trunk in a working day to keep that great bulk going; and if he was (as he largely is) a browser, his mouth wouldn't need to reach the ground, anyway. When he came to drinking water, he could have waded in, or knelt down—it would have been no more inconvenient than the giraffe's straddle.

Richard Carrington reports: 'In Africa few rivers are sufficiently deep to cause so large an animal as an elephant to need to swim, and it is customary for the migrating herds to ford them. Sometimes the water will cover the animals completely; they will then walk across the river bed with only the tip of their trunks showing like periscopes above the surface.'

Now, that would have been a sensible reason for beginning to grow a trunk, especially back in the Eocene, when much of North Africa was still covered with the waters of the ancient Sea of Tethys. A snorkel makes sense. Moeritherium, a prehistoric elephant of the Saharan swamps, had already moved his eyes and ears to a high point on his head. If one of his kin left the mud and took up a littoral life on the shores of Tethys, it would help to have his nostrils on a stalk (like gorgops' eyes) as the water rose higher around him, especially since he was the wrong shape for getting up on his hind legs.

However, if we are to establish that he went marine, it's not enough to have him paddling in the shallows, even with the aid of his snorkel. He would have to be able to swim.

George P. Sanderson: 'Full-grown elephants swim perhaps better than any other land animals. A batch of seventy-nine that I despatched from Dacca to Barrackpur, near Calcutta, had the Ganges and several of its large tidal branches to cross. In the longest swim they were six hours without touching the bottom; after a rest on a sandbank, they completed the swim in three more; not one was lost. I have heard of more remarkable swims than this.'

Perhaps this was under duress, although Carrington comments that 'elephants will sometimes go swimming for the sheer joy of being in the water. The calves dash about on the shore squealing with excitement and pushing each other into the water.' Just like the young of homo sapiens.

Still, the Ganges, tidal or not, is still only a river; it's a different matter to strike out into the open sea. But Carrington relates an instance of this, too: 'More recently a case has been recorded by Lieutenant-Colonel Williams of an elephant that went for a two-hundred-mile island-hopping jaunt in the Bay of Bengal. The animal took twelve years to complete the journey and some of the hops from island to island were across at least a mile of open ocean.'

Let's look at the animal even more closely. If he really lived in the salt sea, it would be logical to suppose that, like ourselves, he might have attempted to deal with his salt-balance problem by special activities of his lachrymal glands.

Darwin has a paragraph on this: 'The Indian elephant is known sometimes to weep. Sir E. Tennent, in describing those which he saw captured and bound in Ceylon, says they "lay motionless on the ground, with no other indication of suffering than the tears which suffused their eyes and flowed incessantly". Speaking of another elephant he says: "When overpowered and made fast, his grief was most affecting; his violence sank to utter prostration, and he lay on the ground, uttering choking cries, with tears dribbling down his cheeks". In the Zoological Gardens the keeper of the Indian elephants positively asserts that he has several times seen tears rolling down the face of the old female, when distressed by the removal of the young one.'

Darwin was writing a long time ago, and since I had found no reference to this in more recent elephant books I thought I'd better do a bit of checking. I visited the library of the Zoological Society of London and was shaken to discover that all this copious activity of the elephant's lachrymal glands could not take place for the simple reason that it allegedly hasn't got any. A monograph on *The Elephant in East Central Africa* affirms: 'There is no true lachrymal apparatus, but a Harderian gland, associated with the nictitating membrane, lies between the rectus medialis and the medial wall of the orbit. Its duct opens on the surface of the nictitating membrane.'

Yet when I walked across the road into the zoo and asked the elephant keeper about it I got the same answer that Darwin had got long long ago: 'Yes, I've seen them cry.' I asked him on what occasions. 'Well,' he said, 'I suppose it's if anything's upset them, yes, they'll cry then. Or at least,' he cautiously qualified it, 'at least, they'll shed tears.'

I believe they do shed tears. I think that we use our lachrymal glands, the seagull uses his nasal glands, and the elephant uses his Harderian glands; but I think we all shed tears.

I won't ask you to look at an elephant's feet, because I tried that myself, and failed to find anything marine about them. However, while I was in the zoo library and glancing through C. Harrison Matthews' *The Life of Mammals* I encountered the following statement which you may interpret in any way you like. The elephant's foot, it appears, has 'five digits in each foot, united by a web'.

I will ask you to look at the fit of his skin, and that of another naked pachyderm, the rhinoceros. (The rhino was a wallower, and the Indian rhino is still pretty aquatic. It feeds on water plants, and Paul Steineman of the Basel Zoo, the first to breed these animals in captivity, says: 'They spend so much time in the water in summer that one visitor asked me in all seriousness whether they lived on fish.')

Where is the adaptiveness in all that bagginess of skin, and those deep deep creases? They only harbour ticks and parasites, so that the rhino has had to contract a symbiotic relationship with a small bird who travels around with him to fish them out of his crevices. However, if you could find a way of inflating that skin with a bicycle pump, to discover what shape the rhino and the elephant used to be when that slack was all filled up with subcutaneous fat, you'd be getting much nearer to the roly-poly shape of a seal or a dugong. Naturally, a good deal of that fat had to be shed again as they began to spend more time on tropical land masses, so naturally their hide doesn't fit too well nowadays. Any successful member of a weight watchers' club will tell you the same about his trousers.

Then, again, if the elephant went marine, it would be logical to suppose that he would have learned to bring the emission of vocal noises under the control of his will. He did, too. If you tell him to trumpet, he will trumpet.

If he did indeed survive the experience of completely changing his ecological habitat twice in succession, this would give as violent and salutary a shake-up to his metabolism and modes of perception as it gave to ours, and his brain, like ours, would need to become more complex, readier to respond to unprecedented problems with flexible solutions. And if you compare the elephant's intelligence with that of its nearest zoological relatives you will find a gap quite as stupendous as that between the apes and ourselves.

If the elephant really went marine, it might be logical to wonder whether, like ourselves, it underwent the usual sexual sea-change, moving the vagina around to the ventral side. If you look at a cross section of female elephant, you will at first think: 'No. Nothing of the kind has happened.' Because there is the vagina, just where it is in a monkey, or a cat, high up under the tail and lying parallel with the spine. But then you will notice something utterly fantastic. There is no exit under the tail. (I suppose an andocentric would have said: 'No entry.') From the posterior end of the vagina comes a tube called, quaintly, a 'vestibule', and this runs within the body wall all the way down the vast hinder end of the animal and curves forward under its belly and finally emerges at a point roughly equivalent to that of the male's penis. The exterior female genital organ of the elephant is even more ventral than yours or mine. It's as ventral as a dolphin's— even though after millions of years of reversion to a quadrupedal landbound mode of progress, the interior arrangements have returned to their former situation.

Carrington: 'This unusual distribution of the sexual organs led the ancient naturalists to believe that elephants must copulate face to face in the conventional human position, and this was regarded as additional proof of the animal's wisdom and intelligence.'

If they ever did this, they don't any longer. He comes up behind her and sinks down very low, almost sitting on the ground, until he has made connection, and then rises to a more nearly normal mounting posture.

But if anyone can think of a good alternative evolutionary reason why the elephant developed this long long tube, necessitating a steep uphill journey of several feet for every sperm before it reaches its destination (*and* she doesn't lie on her back till it gets there, either), I'll be very interested to hear about it.

Now back to *homo sapiens*. In ourselves, too, scores of unexplained characteristics still persist. Some are too trivial to have commanded much attention; others are so basic that we seldom realize how abnormal they are.

Why, for instance, are there so few female counterparts of Peeping Toms? Why no chartered busloads of emancipated women making a beeline for the big city to sit through an evening of all-male striptease? The covers of magazines selling to men frequently feature a plasticized portrait of a beautiful nude girl. This doesn't mean that its readers are dirty old men: it is perfectly right and natural for a man to be pleasurably stimulated by the sight of a nude figure of the opposite sex.

But what do the covers of women's magazines most frequently feature? Why, *another* beautiful girl! It is more often her head than her body, and she is more often clothed than nude; but the publishers are not in business for the fun of it, and if male nudes would sell more copies to women than pictures of girls in pretty dresses, they would be printing male nudes. And they don't. Why is this? And why is it felt to be quite reasonable to explain a female horror of snakes by saying they are phallic symbols?

When you think about it, this is very strange. To many non-human female primates the phallus is a beautiful sight. A mandrill, for instance, will sit upright and flash at a female an erect penis permanently spotlighted in brilliant red and enhanced by scrotal patches of vivid blue and she will find the sight entrancing.

In fact, in every primate species, where there is visual adornment, it attaches to the male, just as it does in birds. It is the male peacock and the male lyre bird that have the most splendid tails; the male hamadryad baboon that boasts the flowing mane; the male proboscis monkey whose imperial nose has burgeoned above and beyond the call of common sense. The females in all these species are comparatively drab creatures whose role is to gaze and admire. Even the hominids—some of them, at least—started out along the same path by sporting gorgeous beards. But at some point, for some reason, in *homo sapiens* alone, these roles became a little blurred so that today we tend to assume that the female of our species is more gazed upon than gazing.

Despite millions of years of disillusion, man has not totally lost his conviction that like the mandrill, he has an irresistibly beautiful sexual appendage, even without the primary colours. He admires it himself; if it is large enough, his mates in the locker room admire it. But a small doubt has crept into his mind. Today when men speculate about women's reactions to a mandrill-type display of it, they become positively schizophrenic.

At one moment they will speak as though the sight or

the representation of a penis must be 'a treat' for a woman, just as female nudity may be a treat for a man; or they will accept Freud's ludicrous contention that she regards it as the tragedy of her life that she hasn't got one; yet in the next breath they think it quite natural that a snake, because it bears a (pretty remote) resemblance to a penis, is enough, for that reason alone, to give her the screaming abdabs. And the man who ambushes women to indulge in a spot of indecent exposure apparently hopes to elicit not admiration but shock and revulsion; and more often than not he succeeds.

A perfectly honest aesthetic assessment from the average woman, however much tactile delight she may derive from it, would probably be: 'Well, let's face it, it's not very pretty.' However, this explains nothing. It is no less pretty than any other ape's, but in no other ape is the response to it so ambiguous, and it is as true in this context as any other that beauty, or the lack of it, is in the eye of the beholder. No well-adjusted warthog finds anything unpleasing in the appearance of her mate, nor he in hers. It would be violently unnatural if they did. So if men take more visual delight in women than women do in men, it cannot be that there is something wrong about their bodies. There must be something wrong about our eyes.

It is not the eyes themselves: they transmit what they see accurately enough. But in the brain behind them, ancient tracks have been laid down that not only receive visual impressions but dictate irrational and sometimes even violent emotional reactions to them. Thus we may react with fear to a visual image not because it necessarily threatens us now, but because to some distant ancestor it has distasteful or frightening associations. The idea of race memory in the straightforward sense in which Samuel Butler believed in it has been discredited; but, as Haldane himself specifically states: 'An instinct can evolve *as if* memory were inherited.'

One of the most famous illustrations of this mechanism was an experiment carried out by Niko Tinbergen, which showed that reactions of cowering and terror may be produced in turkeys, pheasants, and gray-lag geese, by the passage over their heads of a moving shape like that of a flying predator. The reaction came whether or not the birds had ever had an opportunity to see, or to be alerted (by maternal warning or agitation) against such a predator in real life.

The form and motion of the object were crucial. If the same shape was moved in reverse no panic was aroused, because it then suggested not a hawk with a long tail behind him, but a harmless goose with a long neck in front. Somewhere in the pheasant's head is planted an unshakable conviction that a particular form of visual image falling on the retina of the eye is very nasty, and means trouble.

It is possible that we have similar quasi-race memories imprinted in our own brains, even if we are not always fully conscious of their operation. Some years ago in America a technique was developed by a scientist called Hess for detecting subconscious reactions to visual images, by measuring and recording the involuntary dilation and contraction of the pupils of the eye while looking at pictures of various objects.

It was found that when pictures of smiling faces were shown, the pupils dilated, but confronted by scowling or threatening ones they contracted. On seeing the picture of a gory traffic accident, the pupils dilated widely with shock but then at once sharply contracted because the sight was painful and unpleasant.

An interesting discovery was that these reactions were to an appreciable extent sex-linked. Measured by this process, women showed considerable interest in pictures of babies, as might be expected, and men much less. It was also understandable that the sight of a baby with its mother was even more pleasurable than the sight of a baby in isolation, which in a primitive context would occasion some anxiety.

It would have been less easy to foresee that on this scale men showed a much more intense interest in pictures of landscape than women did. But this differentiation can be paralleled in other species, even down to Tinbergen's seagulls, where the male shows an acute and anxious awareness of the location, extent, and limits of his territory while his consort behaves as if she thought, like Kipling's cat, 'All places are alike to me,' and has to be repeatedly grabbed by a wing and yanked squawking back inside the invisible boundary.

These pupillary reactions are totally independent of and uncontrollable by the more conscious level of the brain. An antipermissive pillar of respectable society may be vocally critical when shown the picture of a billowy nude; he may frown and tut and call it vulgar and disgusting, but the pupils of his eyes are silently emitting the wildest of wolf whistles. And I believe these phenomena may offer as valuable clues to our evolutionary history as the structure of a jawbone or the ontogeny of a foetus.

If you show a woman a picture of a male nude her pupils will also dilate, especially if he's young and dishy. But if you call in the aid of another piece of equipment designed to measure where her gaze rests, you'll find that a woman shown a picture of a man looks long at his head, and his face, and his torso, and his arms, as well as his abdomen, while a man shown a picture of a woman frequently doesn't even bother to glance above the neck. At some point or other they are clearly starting from a different premise.

So if her final assessment of the lower half of a male fully frontal and rampant is that it is an interesting but not a prepossessing sight, and certainly not one that she'd either lurk or pay to get a look at, we cannot write this off as a cultural reaction. It is possible that, unlike the mandrill, she passed through a prolonged evolutionary stage in which its associations were more disturbing than delightful. Even while her mind and body may be saying 'Nice', something behind the back of her retina says 'Nasty'. Which is perhaps why we get the odd phenomenon reported by Kinsey, that a man prefers to make love with the light on, but a woman prefers it in the dark. That way she can react all in one piece.

While we're at it let's speculate about the spider and the snake which give so many of us atavistic creeps. The sight of a tarantula can shrivel up our pupils like nobody's business. It's not sufficient explanation to say that some species of spiders are poisonous. Spiders that sting people are infinitely rarer than flying insects that sting people; but though we regard bees with caution and respect, not many people respond to them with horror.

Desmond Morris notes a marked sex difference in the response to spiders, and decides that they must be yet another symbol. 'The only clue here,' he writes, 'is the repeated female response to spiders being nasty hairy things. Puberty is, of course, the stage when tufts of body hair are beginning to sprout on both boys and girls. To children, body hairiness must appear as an essentially masculine character. . . .' I don't find this totally convincing, even though a survey showed that feminine queasiness about spiders increases at puberty. I don't believe it because the same girl who shrieks at a spider will drool over a woolly-bear caterpillar, and carry it home in her hands.

I think, as with the pheasants, it is the outline and the motion she reacts to; and I think the anti-spider pattern was laid down like other ocular reactions on the shores of that ancestral beach, when almost the only thing nimble enough and fearless enough to constitute any danger to her or her helpless infant was that other low-bodied long legged scuttling arthropod, the crab; and it's only to be expected that she gets more worried about arthropods when she reaches an age to bear babies who can't run away. Perhaps some crabs were bigger and fiercer in those days; but anyway if you're naked too long on a tropical beach the little ones no bigger than spiders can be by far the worst, for some of them will get in under your skin and settle there as parasites and eat your arms and legs into horrible sores.

Our phobia about snakes has usually been explained as something we brought down with us from the trees, because if you show a snake to a chimpanzee in a zoo he will panic.

I'm not too sure his horror of it is of quite the same nature as ours, though. Vernon and Frances Reynolds have pointed out: 'There is evidence that wild chimpanzees do not show the panic reactions to snakes that are shown by captive chimpanzees. Adriaan Kortlandt fastened a large living viper on the track of a group and the chimpanzees merely stepped aside a few feet.' We are naturally unwilling to believe that they are less irrational than we are in this respect; but in fact the snake thing, like the spider thing, appears to be a human speciality.

Perhaps it's just coincidence that the other major threats to the hominid in his shallows may well have assumed precisely this form. This would apply not only to the sea-snakes—many of which are extremely poisonous and far more vicious than their land-based relatives—but also to the eels which lay anchored under the rocks where he did his diving. Their muscles are far more powerful than his and their teeth never let go; if one of them made an underwater grab at his finger or toe, and he didn't have a pebble tool sharp enough to saw off either his finger or the eel's head, then he would certainly never breathe air again. Agreed, it seems a far-fetched theory. But statistical inquiries have shown that the snake is by an enormous margin the animal that people most loathe (with the spider coming second) and the strangest thing about it is that when the questionnaire is expanded to determine what quality of the creature arouses this disgust, the majority of the snake-haters explain quite readily that they can't stand the creature because it is 'slimy'. Now, an eel is very slimy, but a snake's skin is as dry as a length of sunbaked rope. That slimy snake that haunts our nightmares exists nowhere on God's earth, except in the backwaters of the race memory of homo sapiens.

All this of course is highly speculative, though no more speculative than other attempted explanations of the same phenomena. But if there is any truth in it at all, it is fairly important for us to get these things the right way around in our minds.

Most people writing about sex, and particularly about women's attitudes to it, have tended to assume that there must be an overwhelmingly strong and unequivocal biological predisposition to approach it with joy and desire, and that any hesitancy or misgiving must be the result of prudery and artificial inhibitions, and therefore should be countered with reproach and derision; or that it might arise from perversion or seeing something nasty in the woodshed, to which the proper response is: 'You're sick.'

It is quite possible, as with the fear of the innocent grass snake, that the truth is the other way around: that there is an *innate* predisposition to anxiety, which can be overcome by reason and experience and assurances from the tribe that this is a Good Thing, but which any alarming encounter will powerfully reinforce. It is the fear which is subcortical; it is the conscious mind which learns to discount it as atavistic and maladaptive, like the fear of thunderstorms.

Most women have at some time felt indignant—quite rightly—about men's feelings of revulsion towards them, their horror of the 'uncleanliness' of menstruation, the custom of churching after childbirth, and the rest of the barbarous nonsense. But men have had their share of being called dirty and beastly, too; there are motes in our eyes as well as in theirs. It will be a good thing when we can all get rid of them. It will be an equally good thing if we don't expect miracles of one another in this respect; if we will tolerate in them, and they will tolerate in us, the fact that some elements in this mutual queasiness have roots that go back not, as Freud thought, twenty years to an individual infancy, but fifteen million years— to the infancy of the race.

Eight

Man the Hunter

We have now arrived in the Pleistocene. The hominids are moving up the rivers and settling on the shores of the new inland lakes. They are so accustomed after ten million years to using pebbles for tools that they use river pebbles, and ultimately (in inland areas where there are no pebbles at all) the nearest thing to pebbles they can find, such as flintstones. They are so accustomed to living in caves that since caves inland are harder to find they tend to settle in places where they can find them, and leave their bones and their relics there: hence 'cavemen'.

Where they had been accustomed to covering the floor of the cave with dried seaweed, they now covered it with straw or furs of animals. Archaeologists have found one ancient cave in Nice, France, where relics of both traditions survive. They unearthed in the cave the remains of the place where the fire burned, and around it a pattern of tiny shells from the kind of molluscs that cling to the strands of seaweed, and just above that a number of claws (but no bones) from the skins of animals. Clearly it was not a gigantic step they had to take when the Pleistocene rains came down; it was not much more than a change of locale.

But the gap was over. Dramatically, their bones appeared again on the plains, where for millions of years no trace of them had been left. As Ardrey puts it: 'We came marching back from wherever we had been.' And we came back sea-changed and different: upright, bare-skinned, omnivorous, tool-using, in the first stage of recovering from the biological emergency, and in the first stages of true verbal communication.

I propose to examine in detail four facets of the androcentric legend concerning the major developments of this stage of our history. One is a rather romantic minority report by Robert Ardrey, who suggests that the seeds of our war-torn society were sworn at this time, because there were two speices of Australopithecines, one peaceful and one belligerent, and the belligerent one wiped out the peaceful one and was the ancestor of the bloodthirsty human race.

The other three are far more important, because they are almost universally accepted. Specialists may have reservations about them, but their reservations have not for the most part percolated into the public consciousness: the man in the street swallows them whole.

The first myth is that man at this time became a hunting carnivore, and that woman, being a non-hunter, stayed at home with her offspring waiting for her mate to bring home the sustenance without which none of them would have survived.

The second myth is that, as a consequence of this, woman already at this early date dwindled into a housewife, contributing nothing to human culture, while the males laid down all the foundations of technology and art.

The third myth is that during this time the human race became pair-bonded, because man the hunter needed a peaceful mind based on sexual monopoly, so that he kept a woman in his cave, feeding her and her children in return for sexual favours, and thus he originated the monogamous nuclear family—one man and one woman and their offspring.

First, then, the Australopithecines. There is no doubt that more than one species of manlike creature reappeared in Pleistocene Africa. Skulls and teeth and femurs and relics have been dug up in various parts of Africa (almost exclusively on the sites of ancient rivers and the shores of ancient lakes). Two species in particular have been recognized, which we may group under the generalized headings of *Australopithecus robustus* and *Australopithecus africanus*. *Africanus* was considerably smaller; judging by his teeth he was more carnivorous; his cranium was more like our own. *Robustus* was larger and clumsier, with a crest on his thick skull like a gorilla; he had large grinding molars with thick enamel suitable for chewing vegetable fibre, and they have been found with abrasions, suggesting that possibly he was in the habit of digging up roots and failing to clean all the sand and grit off them before eating them.

Not everyone recognizes the division into two species, or subspecies; but these general types do polarize some of the differences between the hominids, and a great deal of argument has been expended as to which of them (if either) fathered the human race. *Robustus* has the uglier, more brutal-looking cranium, and *africanus* has his caves heaped high with the bones of slaughtered animals, bringing him nearer to the popular Tarzan figure. It is tempting therefore to regard him as of 'higher' evolutionary rank, but unfortunately his bones date from even earlier than *robustus*'s.

Robert Ardrey solves the problem neatly. There is never much doubt which side of the fence he will come down on. He casts the vegetarian *robustus* as Abel, and *africanus* (who was given to slaughtering his own kind as well as the beasts of the field) as Cain. Somewhere, sometimes, he postulates: 'Cain met Abel, and slew him, and made his weapons thenceforth of quartzite and lava, and fathered the human race.' So we must be descended from *africanus*, whose remains allegedly afford 'a positive demonstration that the first recognizable human assertion had been the capacity for murder.' And anyone who doesn't want to believe it is a starry-eyed liberal.

In the years since *African Genesis* was written, the facts have become a little clearer. It now seems plain that the war, or duel, between this particular pair could never have taken place, because nowhere in the whole of Africa have relics of the two species been found to co-exist. *Robustus*, it appears, must have died not by murder but of natural causes, as countless other species have done before and since.

As time goes by he becomes less and less likely a contender for the title of ancestor to *homo sapiens*. The theory that he invented the first flaked pebble tools was not upheld by later researches—the tools are now usually attributed to *Homo habilis*—and the latest fossil discoveries have given rise to the suggestion that he was not even a proper biped, but walked on his knuckles, like a gorilla.

Perhaps these animals were descended from a strain of *Dryopithecus* which had found some relatively un-arid corner of Africa and managed to hang on there in small numbers throughout the Pliocene, and never made it to to the sea at all. When the temperate times returned they began to spread and multiply, but lacking the new skills and adaptability of their streamlined aquatic cousin *africanus*, they never quite qualified for permanent survival.

But what I am most concerned to combat in the Ardrey thesis is the assumption that if two populations of near-human primates had encountered one another in those primitive days, the inevitable outcome would have been genocide.

Let's go back a bit, and talk about the process of speciation.

As Darwin observed when he studied the fauna of the Galápagos islands, all that is needed to produce two species of a bird or an animal where only one existed before is to segregate them by some barrier (usually geographical, such as a stretch of ocean) into two groups which are thus prevented from interbreeding. If the separation continues for a long period the two populations will begin to diverge from one another, in appearance or behaviour or both; if it continues long enough the divergence will become so extreme that even if the two populations are then enabled to intermingle, they will not interbreed.

Now, if we go back to the Miocene and think about little *Dryopithecus* at the very beginning of the story, we may assume that she had more or less free range through all the forests and jungles that covered Africa. We know too that when extremes of temperature, whether of heat or cold, strike at a continent, the areas farthest inland suffer most while the coastal areas remain comparatively temperate. It is possible, then, when Pliocene Africa began to bake and crack and the forests to dwindle, that outside the equatorial belt and the highland enclaves where the emergent gorillas hid out, the forest would tend to die from the centre outward to the sea; so that the ancestral ape, retreating with the forests, would not be backing towards the ocean in one straight line. Small groups of these animals would be driven out, as it were, radially from the centre of the circumference, and would reach the safety of the sea at different points around the coast of Africa. There they would live, possibly for very long periods quite out of touch with one another, but inevitably moulded in roughly parallel directions by the influences of their new environment.

However, some morphological divergences would take place, facilitated by this dispersal. Any minor mutation would have a chance to establish itself in a small isolated group, where in a larger population it would be more likely to be swamped. There can be no doubt that the race manifested even wider ranges of variation then than it does now.

Some of the variations died out. Some survive only in populations that have remained undisturbed perhaps from the very beginning. The Bushmen of Africa, for instance, display physical peculiarities, like steatopygia, which are not found in other races. Some of their unique attributes, although they have now been driven to take refuge in some of the most arid areas on earth, may have originated in the aquatic era. For instance, in the days of the biological emergency it would have been advantageous not to have a difficult situation made more hazardous by the uncertainty of attaining and holding an erection: the male Bushman's penis is permanently semi-erect, from the hour of his birth until he dies. The women are born with a natural covering over their genitals called the *tablier egyptien*. It could have been an incipient example of aquatic streamlining such as that which encloses the dolphin's nipples in a retractable sheath of skin.

Differences in size would have emerged also. Dispersal of a species almost invariably results in variations of size; with most warmblooded species, latitude alone will bring this about. The nearer poleward they live the bigger they become. Puffins increase in size by over 1 per cent for each degree of north latitude, until Spitzbergen puffins are nearly twice the size of Brittany ones. In southern latitudes the same is true of penguins.

The term sometimes used by biologists to describe such an isolated breeding population is 'deme'. The question is, what happened when the scattered demes of the aquatic ape got together again? It may have happened now and again during random migrations up and down the coast; but the chances of such encounters would be vastly increased when the Pleistocene fertility opened up the interior to them.

They would meet as strangers. Perhaps, as we have seen, they would differ markedly from one another in appearance. Perhaps the males of one of the demes would attack, and there would be a fight. It is not such a foregone conclusion as some writers would suggest, for different species of primates often dwell together at peace in the same stretch of forest; but even if they fought, the outcome would hardly be genocide.

It is unlikely that in such a scuffle all the unresisting females and young would also be murdered. And if they were left alive, we can be pretty certain that the widows and orphans, belonging as they did to a compulsively social species, would tag along after the victors rather than opt for dispersal and isolation, even though the victors were by their standards too tall or too short or too black or too long-haired, or ate strange food or made strange noises.

It is at this point that the Galápagos-type mechanism breaks down. In most other species a wild divergence in appearance or behaviour constitutes a gap too great to cross. Imagine, for instance, a species of birds which had been separated on two islands until there was a divergence of, say, 20 per cent in the size of the two populations. By the operation of natural selection, those demes even if reunited would most often not interbreed but remain forever distinct and crystallized into two species—as it were the great auk and the little auk.

And the factor by which natural selection ensures this is the choosiness of the females, who tend to be conservative in their preferences. From Drosophila upward, they tend to reject males who look or behave differently from the norm. And from Drosophila upward (with one sole exception) they are the final arbiters. If they kick and dodge and extrude their ovipositors the wedding is simply not on.

But the hominid's female, when she returned to the land, was no longer either the initiator or the final arbiter in the selection process. Widows and orphans who accompanied a strange victorious troop would have no chance to bleat about the purity of the deme. Perhaps, indeed, they had begun to lose the instinct to do so, and to participate to some extent in the male's 'vagueness of aim'. In any case, they were probably regarded as loot. So even a variation in magnitude—say a foot of body height—would have been no bar to interbreeding then, and never has been since. The only bars are geographical and cultural.

So I cannot swallow Ardrey's assumption that since there were large Australopithecines and small ones, we must conclude that they fatally clashed, and we must further decide which stock we are descended from, because one must have exterminated the other.

Once again he has forgotten the females were there. It is like saying, because William the Conqueror clashed with King Harold, we have to decide whether the English are descended from the Normans or the Saxons.

The variability, the versatility, the viability, of humanity derives in no small part from the fact that it has remained biologically one species; that however often new barriers have intervened—geographical, social, or any other—however often human populations have hived off and diverged in morphology or colour or culture, once contact is reestablished the streams always begin to flow together again and enrich one another, and enhance the evolutionary potential of the race as a whole.

If we owe this unique and infinitely fruitful facet of human biology to the ousting of the female from her position of what Ardrey calls the 'sexual specialist', then the biological emergency, however painful its repercussions, may have been one of the most creative things that ever happened to us. If it had not happened, then since men have spread over the earth even more widely than baboons, there would today be more species of Homo than there are of Papio. As it is, however hard the racists try to blind themselves to the fact, there is only one.

There is only one, and—until the arrival of cultural mutants such as George Bernard Shaw—he has for millions of years been a carnivore. This evolution of a meat-eating primate is sometimes treated as a new and unique phenomenon. It is quite true that primates on the whole overwhelmingly favour a vegetarian diet. In an arboreal habitat vegetation is the most abundant type of food; and they have become in some ways specialized for it. The sweetness of fruit sugar, for instance, is attractive to them, whereas some of the true carnivores like the cat (though oddly enough not the dog) have no taste buds capable of perceiving sweetness. If you try to please your cat by sprinkling sugar on her bread and milk, the effort will be as wasted on her as the colour scheme you've chosen for her sleeping basket, for sweetness is as far outside her ken as yellow or blue.

But it is not true that, as Ardrey stated in *African Genesis*, 'Any mutation in a primate species making possible the digestion of meat must be a transformation of revolutionary genetic proportions.' It may well have accorded with the facts as far as they were known when he was writing, for primate studies are progressing so fast that any book dealing with them (this one, too) is liable to get shot full of holes a few months later by new discoveries in the field.

Anyway, we now know that anthropoids in the wild often eat the flesh of animals and fish, and not infrequently are willing to kill to obtain it. The crab-eating macaque gets most of its protein by catching and eating crustaceans. All baboons will eat small animals and fledgling birds; some will kill and eat guinea fowl; some will deliberately chase, kill, and eat hares, young gazelles, or even vervet monkeys. Chimpanzees will occasionally kill and eat young bush buck or bush pig. Jane Goodall saw one grab a colobus monkey, kill it by breaking its neck, and then eat it.

None of these animals makes a regular habit of this sort of behaviour, but as a way of adding variety to their diet, meat obviously intrigues them. Other chimpanzees will cluster around the killer and hold out their hands for a share, like children asking G.I.'s for gum or chocolate, though they don't display this behaviour where vegetable food is concerned, because it is all around them for the taking.

It is quite clear, then, that the hominid was not the first or only primate butcher; nor did he need any 'revolutionary genetic transformation' to enable him to digest meat, because a great number of his primate cousins can already do it with the greatest of ease.

Possibly some of the scattered hominid populations did not alter their primarily vegetarian habits until after they had returned to the land. There would be nothing particularly surprising in this. All the coastal demes of the aquatic ape would need to learn new eating habits. Just as some of them graduated from eating insects to eating shrimps and later to sea mammals, so others in the same plight may have graduated instead from land vegetation to sea vegetation—from leaves to seaweed. It was always something of a toss-up which way they would go. Seaweed is highly nutritious food. Steller's sea cow lived on nothing else and sustained a bulk many times greater than ours. It is palatable, too, although it has recently gone out of fashion. There have always been, and still are, regions where, in such forms as dulse and laver bread, it has been considered a great delicacy.

The chances are that the herbivore/carnivore dichotomy was never as clear-cut as we sometimes imagine. Dr. Leakey's *Zinjanthropus*, for instance, had vegetarian-type molars, but the bones of small animals lie around him in his grave. Like the baboon and the chimpanzee, he wasn't doctrinaire about what he ate.

The converse of this is even more significant. The hominids who had become meat-eaters weren't doctrinaire about what they ate, either. We are most of us meat-eaters ourselves: yet ounce for ounce meat constitutes only a small part of our total consumption. The same was almost certainly true of our hunting ancestors. There

seems to be no reason to doubt Bartholomew and Birdsell's conclusion that 'Like most present-day hunting and
collecting peoples the Australopithecines probably used
plants as their major source of food' and lived by
eating 'eggs, fish, crustaceans, insects, small mammals, reptiles, scavenging from larger carnivores, berries, fruits,
nuts, roots, tubers, and fungi'.

When conditions were bad there would be less variety.
A cave in the Zagros Mountains in Iran excavated by
Ralph Solecki in 1956 proved to have been the home of
man more or less continuously for thousands of years,
and in layer 2—corresponding not to Australopithecus but
to the Middle Stone Age of *homo sapiens* proper—he
seemed to have been going through one of his lean times,
with no trace of agricultural products or domestic animals,
and 'judging from the great number of snail shells found,
it is likely that snails were a main staple of their diet'.
Early man, however heroically he depicts himself on his
cave walls, was not always roasting haunches of mammoth
and wildebeest.

I make this point not to belittle his achievement. At
his best I am sure he was magnificent—his eye for spoor
not much inferior to the felid's nose, his gait untiring, his
courage high, his aim deadly, his ingenuity unparalleled,
and his welcome when he brought home the bacon warm
and uproarious. It is all perfectly true, about his skill and
his daring and his weapons and his intelligence.

Yet the balance does need to be redressed. Because when
a Tarzanist says to himself 'carnivore', he thinks of a wolf,
and he thinks: 'Yes, of course, that's what happened to our
society. The woman stayed home in the cave with her
litter, like the female wolf with new pups who can't
join in the hunt. They stayed there until he came home
with the gravy, all of them just lying around waiting to
help him eat it.' And if the Tarzanist lives in a society be-
devilled with momism and alimony and executive ulcers,

the picture will crystallize in his mind and carry subconscious corollaries of 'and that's all she's done ever since' and 'all *he* got out of it was a chance to sleep with her when the hunt was over, so my God I'll bet she had to be pretty good at that if she wanted her slice of venison'.

Bernard Shaw put into the mouth of Cain, in *Back to Methuselah*, the essence of the romanticist's version of the hunter's home life:

'I will hunt: I will fight and strive to the very bursting of my sinews. When I have slain the boar at the risk of my life, I will throw it to the woman to cook, and give her a morsel of it for her pains. She shall have no other food, and that will make her my slave. And the man that slays me shall have her for his booty. Man shall be the master of Woman, not her baby and her drudge.... Only when [a man] has fought, when he has faced terror and death, when he has striven to the spending of the last rally of his strength, can he know what it is to rest in love in the arms of a woman.'

Eve's answer is intended to present the other side of the coin:

'You her master! You are more her slave than Adam's ox or your own sheep dog. You slay the tiger at the risk of your life; but who gets the striped skin you have run the risk for? She takes it to lie on, and flings you the carrion flesh you cannot eat. You fight because you think that your fighting makes her admire and desire you. Fool: she makes you fight because you bring her the ornaments and treasures of those you have slain. What are you, you poor slave of a painted face and a bundle of skunk's fur? ... You are to other men what the stoat is to the rabbit; and she is to you what the leech is to the stoat.'

All this is fine dialectic, and a beautiful encapsulation of the pros and cons of the myth of Man the Carnivore, a myth still alive and kicking in the collective subconscious of all the huskier male-bonded groups.

'The female,' says Desmond Morris, 'had to stay put and mind the babies'. It would appear that this was all she had to do all day long. And, of course, limber up for a possibly 'bizarre elaboration of sexual performance' when the hunter came home, to keep the pair bond well cemented.

S. L. Washburn, in a much more rigorously scientific context (a symposium on Man the Hunter in 1968), draws a very similar moral: 'Hunting changed the role of the adult male in the group. Since sharing the kill is normal behaviour for many carnivores, *the economic responsibility of adult males* and the practice of sharing food in the group probably result from being carnivorous.' (My italics.) At this point in prehistory, runs the implication, females turned into dependants and consumers.

It is high time this whole legend was exploded, because it is not just a myth pure and simple: it is a political myth. It is used to bolster up with pseudo-history and pseudo-anthropology the belief that it is 'against nature' for women to play a part in economic life; that 'from time immemorial' men have said 'she shall have no other food and that will make her my slave'; and that we are descended from females whose sole function was to placate the hunters and keep them happy and mind the babies.

It was never really like that; and among the surviving hunting communities it is not like that now. For one thing, if the women really behaved like that they would starve. Richard B. Lee, who studied the economic life of the !Kung Bushmen of Botswana for fifteen months, reports:

'It is not unusual for a man to hunt avidly for a week and then do no hunting at all for two or three weeks. Since hunting is an unpredictable business and subject to magical control, hunters sometimes experience a run of bad luck and stop hunting for a month or longer.

During these periods, visiting, entertaining, and especially dancing are the primary activities of men.'

Laurens van der Post, also in the Kalahari, fills in on the activities of the women and children:

'Daily the young women and children went out with their grubbing sticks to look for food in the sands of the desert. Whenever I accompanied them the intelligence, diligence, and speed with which they harvested the earth never ceased to astonish me. A tiny leaf almost invisible in grass and thorn just above the surface of the red sand would cause them to kneel down and grub deftly with their wooden digging sticks to produce what I, in my ignorance of Kalahari botany, called wild carrots, potatoes, leeks, turnips, sweet potatoes, and artichokes. One of their greatest delicacies was a ground nut which, when roasted on their fires, would eliminate all rivals from cocktail counters. And, of course, they loved the wild tsamma melon in all forms, and highly prized the eland cucumber.... All this was achieved at the worst time of the year. I longed to see the riches that could be garnered in the full harvest of summer....'

The children, too:

One night 'the bush and plain was just beginning to resound with the call of nightjars, the melancholy cry of carrion birds, and the mournful bark of jackals.... We overtook a brave little procession of three or four children all up to their ears in thorn and grass. A little boy, grubbing stick in hand, led the procession with a bundle full of roots, tubers, caterpillars, and succulent grubs in his hand. A small girl followed with a bundle of wild and sun-dried berries and rare ground nuts.... The youngest of all carried a large tortoise in a hand held level with his shoulder, and he was breathless with the conflicting efforts of supporting it and keeping up with his elders in front.'

Agreed, this writer loves the Bushmen very fiercely, and the love colours his writing; some people find this slightly embarrassing and may therefore be less ready to accept his facts. So for a purely objective and statistical evaluation, back to Richard Lee:

'Vegetable foods comprise from 60 to 80 per cent of the total diet by weight, and collecting involves two or three days of work per woman per week. The men are conscientious but not particularly successful hunters; although men's and women's work input is roughly equivalent in terms of man-day of effort, the women provide two to three times as much food by weight as the men.' Outside the Arctic this 60 to 80 per cent holds true for *all hunting-gathering groups studied to date*.

(The Eskimos, Aleuts, and other Arctic tribes have a different pattern simply because for much of the year no vegetable food is available, but these communities cannot have much relevance to our evolutionary story, since Australopithecus was an African.)

This then is the true picture of the life of the hunting primate. It's a far cry from the legend of the male bursting his sinews while his mate lolls about on a heap of furs until she and her voracious brood can freeload on the fruits of his labour. If he carries his weapon wherever he goes, so she carries her stamping block for pounding tsamma seeds and protein-rich mangongo nuts into the staple item of their diet.

Of hunting-gathering peoples in most parts of the world, anthropologists in describing their diet tend to use such phrases as: 'Meat is an important prestige food ...' 're-garded as a special treat...' 'a break from the routine of vegetable food....'

Although it is not to be expected that vegetable foods would leave archaeological deposits in the soil to the same extent as the bones of slaughtered animals, nevertheless where the soil conditions have favoured preservation, Mesolithic and Paleolithic sites have yielded evidence, in the form of seeds, nutshells, grinding stones, and digging sticks, confirming that the pattern has not changed very much. Even the hard evidence of human dentition—the vanishing canine teeth and other changes from the anthropoid norm—suggests strongly to such specialists as C. J. Jolly that the characteristic teeth of homo sapiens evolved not to tear flesh from bones, but primarily to facilitate the chewing of grain.

On all counts, then, the idea that in the hunting stage of his development the male took over 'economic responsibility' will simply not hold water. All that happened was that man became the meatwinner and woman the breadwinner.

Yet the myth of the non-productive primitive female persists. I fear we shall never eradicate it from human— especially male—consciousness unless we introduce a government order, like the health-hazard warning on cigarette packets, ruling that every publication dealing with this subject must attach an erratum sticker to the flyleaf: 'Re homo sapiens: For carnivore, read omnivore. For carnivorous, read omnivorous.'

One momentous consequence resulted from the new behaviour patterns of the breadwinning female and her hunting-butchering mate. They now had more time to think, as well as new problems to think about.

Generally speaking, it is the world's vegetarians who are most likely to find feeding virtually a full-time job. A silkworm, for instance, never has a minute to rest from this chore. A wild elephant will spend most of his waking hours locating and eating the six hundred pounds of adequately varied green fodder he needs every day to keep him in condition; and cattle, when taking a rest from eating grass, merely switch over to regurgitating and reprocessing it. By contrast, a full-fed lion or vulture or boa constrictor can bask peacefully in the sun for a whole morning or a whole day, or a whole week, doing nothing but digest.

As vegetarians go, the anthropoid apes already had it fairly good—a gorilla in a lush piece of jungle doesn't need to spend all day eating, and the mid-day siesta is standard practice among primates.

However, the new mixed hunting-gathering economy improved the allowance of leisure still further. Marshall Sahlins aroused some controversy a few years ago by challenging the view of primitive man as leading a gruelling hand-to-mouth existence haunted by the spectre of starvation; he referred to this stage of our development as 'the original affluent society', defining an affluent society as one in which all the people's wants are easily satisfied.

Detailed reports by anthropologists of most hunting-gathering groups other than the Arctic ones confirm this. Frederick D. McCarthy and Margaret McArthur's time-and-motion study of Australian aboriginals reveals that the time spent on procuring and preparing food takes up in one group an average of five hours per day; and in another group less than four. For the !Kung Bushmen he gives two to three days' work per week; James Woodburn for the Hadzas, an average of two hours per day. The standard of living by our criteria is of course not high; but a working week which would drive even an American trade unionist wild with envy enables them to live, rear their children, support non-productive adults, and ensure a varied diet with both calories and proteins well in excess of the dietician's Recommended Daily Allowance.

If the hominids—we can by this time refer to them as *Australopithecines*—lived this sort of existence, it made possible a new move towards human status. A replete lion simply lies around and sleeps in the sun, but a primate, especially a young primate, is a more restless and curious creature. He wouldn't have just sat there all through his three or four weekly days off. He'd have sat there, as it were, whittling, tinkering with something, making something. And this by constant effort and practice would help him to evolve from a small-brained pebble-chipping Australopithecus into homo sapiens, the genuine craftsman.

Here we cannot blame the androcentrics for deducing, as Washburn and Lancaster do, that 'the tools of the hunters include the earliest beautiful manmade objects, the symmetrical bifaces, especially those of the Acheulian tradition'. It was after all the male who had the incentive to fashion stone into functional shapes. The female had uses for stone; but for grinding grain and nuts you basically need only a flat sort of stone to put them on and a round sort of stone to bash them with, and with a little patient searching around you can usually find those ready-made.

Man, however, here begins to emerge as the technologist. The problems of his new trade are exercising his brain; he wants weapons that he can swing and hurl. As Washburn remarks: 'An axe or spear to be used with speed and power is subject to very different technical limitations from those of scrapers and digging sticks, and it may well be that it was the attempt to produce efficient high-speed weapons that first produced beautiful, symmetrical objects.' Tentatively he points even further ahead, from the craftsman to the artist: 'Clearly, the success of tools has exerted a great influence on the evolution of the brain, and has created the skills that make art possible.'

And woman, alas, was never a spear-thrower. Can it be that here, already, we have the beginnings of the dichotomy that explains why there is no recognized female counterpart of Leonardo and Rembrandt and Picasso? It would be saddening to think so.

Let's try to imagine how *she* was spending the leisure hours. When they were on the move, his eyes were constantly busy searching for spoor, droppings, carrion—any signs of prey or predator; her eyes were equally busy looking for desirable berries, leaves, seeds, grubs, or nests with wild bees. His job needed courage, speed, and a weapon; hers needed patience and (now that she was gathering more than she herself wanted to eat) a container. Half a coconut shell wasn't a bad bet. The shell of an ostrich egg was better still because it was easier to remove a controlled amount of it to form an opening. There were times, however, on the plains when there weren't any coconut trees or gourds or ostrich eggs; and when she had nothing better to do she often mooched around looking for something concave and portable and probably circular, since the containers she was used to were circular.

The only things her eyes alighted on that in any way resembled this conformation were the sunbaked footprints left by ungulates in the mud around waterholes. Again and again her eyes would flash the message: 'That's the right shape.' and her brain would reject it because it wasn't portable.

Inevitably one day while the male was chipping flints she would try to dig out one of the footprints whole, and of course it would break. Equally inevitably sooner or later on a long well-fed summer afternoon, having failed again, she would start playing with moister mud and fashion an artificial coconut shell and have to leave it in the baking sunshine while she went to get the supper.

In archaeological terms pottery comes very much later than The Weapon. It is far less durable than stone, and the earliest amateur samples would disintegrate easily and never survive at all. All the same, when we come to consider what was the 'first beautiful symmetrical object' ever fashioned, it must be something of a toss-up whether it was his spearhead or her container. Not that she's ever been credited with the container, anyway. Weapons are the invention of the 'male', the 'hunter', while pottery was, as every schoolboy knows, invented by 'man'.

It is admittedly a generic term and may cover his wife as well; but I can't help feeling that most writers, insofar as they think about it at all, have a vague theory that runs: 'One day he noticed with a secret chuckle that the little woman was wearing herself out trotting to and fro carrying seed home by the handful. He quietly laid aside his beautiful symmetrical weapons and forsook his male-bonded companions for a few weeks while he devoted himself to the problem, and finally invented the pot. He gave her a few prototypes and a crash course of instruction, patted her on the head, and sped away across the savannah to rejoin the hunting party.'

Well, it might have happened like that. No one can prove it didn't; just as no one can prove she didn't take it into her head to say: 'Play quietly among yourselves today, children: I'm busy inventing the bow and arrow for your father.' I would only claim that the second of these tall tales is no more inherently unlikely than the first; for necessity is the mother of invention, and since their economic roles diverged, the container was her necessity, not his. Nobody carries venison home in a jar.

And when we come to prehistoric homo sapiens we have more than probability to guide us. While we can only deduce, however confidently, that it was the males who made prehistoric weapons, the makers of prehistoric pottery left their fingerprints on their handiwork; and the Soviet archaeologist P. N. Tretyakov, among others, has pointed out that the form of the prints indicates quite clearly that they were made by females.

The last major behaviour commonly described as a legacy of the hunting-gathering era is monogamy. Or should I call it 'pair-bonding'? No, I don't think I should. It sounds splendidly scientific, but in fact it enshrines a whole raft of fallacies about the nature of human relationships.

In the first place, homo sapiens is not and never has been a pair-bonding species. Very few species are. The habit of choosing one mate and being faithful until death do you part is indulged in only by a select and strangely assorted brand of creatures which includes the raven, the gibbon, the goose, and the painted shrimp. This characteristic appears to be the only one they all share. They include predators and herbivores; birds and mammals, and crustacea; social species and non-social ones; they inhabit different parts of the planet, different types of environment, even different elements. It would almost seem that whatever fairy godmother had power to bestow the biological gift of lifelong fidelity picked the names of her beneficiaries at random, out of a hat. And there was no piece of paper with our name on it.

If we were endowed with the same biological mating pattern as the goose, there could be no polygamy, no promiscuity, no celibacy, no harems, no group marriages, no trial marriage, and no divorce in any human community in any part of the world. To say 'my ex-wife' would make no more sense than saying 'my ex-sister'. The pair-bonding would come to a man as ineluctably as puberty or death, and he would pair for life with the best unattached female attainable to him during the brief period when he was ripe for such imprinting. If he were in the Vietnam jungle at the time he would pair, *faute de mieux*, with the nearest G.I. who happened to be in the same state of vulnerability; and from then on, with or without any homosexual activity, they would be inseparable as long as they both did live, and Brigitte Bardot herself would be powerless to lure either of them from his allegiance. It is a cast-iron system, rather like Titania's love potion except that there is no known antidote. Quite clearly it is not the system under which human beings operate.

It is of course possible to argue that since the essence of evolution is change and modification we may be moving either out of, or into, a period of biological pair-bonding which is still (or already) partly operative.

The likelihood that we are moving out of it is virtually nil. No species closely related to our own shows any signs of this type of behaviour. The only monogamous ape is the gibbon; and to anyone tempted to equate our family life with that of this distant Asiatic cousin, Washburn and Lancaster direct this word of warning:

'This [gibbon family] group is maintained by extreme territorial behaviour in which no adult male tolerates another, by aggressive females with large canine teeth, and by very low sex drives in the males. The male-female group is the whole society. The gibbon group is based on a different biology from that of the human family and has none of its reciprocal economic functions. Although the kind of social life seen in chimpanzees lacks a family organization, to change it into that of a man would require far less evolution than would be required in the case of the gibbon.'

If on the other hand we postulate that we marry because we are moving *towards* a period of biological pair-bonding, then the movement seems to have been very patchy and sporadic in its operation, and at no time universal, and quite recently showing signs of going into reverse. Such rapid fluctuations are more typical of cultural patterns than of biological ones; which is why I prefer the cultural term 'monogamy' to the pseudo-biological 'pair-bonding'.

The gibbon's 'very low sex drive' is a reminder of a third fallacy: that pair-bonding is based on sexual attraction. Konrad Lorenz, who conducted a classic investigation of the pair-bonding system in geese, states quite categorically: 'The bond that holds a goose pair together for life is the triumph ceremony and not the sexual relations between mates.' Indeed, he adduces some evidence that the tightness of the pair bond in a species is a fairly reliable indicator of its level of aggressiveness towards its own kind. That figures. If you hate the guts of everybody around you, it becomes absolutely necessary to evolve a system that exempts at least one other individual from your general hostility: otherwise the species would never survive.

What we have is not pair-bonding, but a pattern of organization known to scientists as the nuclear family—i.e., Daddy and Mummy and the kids.

Because this is so familiar a feature of our own lives, most people tend to project it rather too indiscriminately on to the lives of our fellow creatures. We tell our children tales about a cozy household of 'Father Bear and Mother Bear and Baby Bear', oblivious of the fact that Father Bear would certainly gobble up Baby Bear on sight, if Mother Bear didn't give the child some rigorous training in shinning up a tree trunk before letting him loose on his own. The Noah's Ark arrangement favoured by many zoos of pairing off one male and one female encourages parents to tell their young: 'There's the daddy elephant and the mummy elephant and the little baby elephant'; 'There's the daddy giraffe and the mummy giraffe and the baby giraffe'; 'There's the daddy monkey and the mummy monkey and . . .' and so *ad infinitum*, as if the nuclear family were a natural feature of the lives of pachyderms, ungulates, primates, etc., etc., to say nothing of cats, dogs, horses, chickens, and sheep.

In the great majority of these cases Daddy's role is purely genetic; his interaction with any individual Mummy is apt to be casual and fleeting, and his individual reaction with his young offspring minimal or non-existent.

The primate group from which we derive is in general no exception to this. Michael Chance and Clifford Jolly in *Social Groups of Monkeys, Apes and Men* point out that most primate males take pains to avoid the vicinity of infants, and this tendency is wholeheartedly endorsed by the females. Among langurs, for instance, a mother with an infant is rarely seen within twenty-five feet of an adult male, and if by any chance he infringes this limit he may be threatened or chased off by the mother.

Daddy Gibbon is the odd man out, in that he does maintain a 'nuclear family' group. But he tolerates the presence of his offspring only as long as they are infants; his wife then co-operates with him in driving them away by harsh treatment as soon as they begin to mature.

There is, however, one group where the males do take a deep interest in the infants. Among baboons and mac-aques a baby is the centre of lively attention from all members of the group. Males as well as females are eager to examine it and hold it in their arms. Adult males are attractive to juveniles and tolerant of their horseplay, and in some species an orphaned infant is as likely to be 'adopted' for grooming and protection by an adult male as by a female.

Yet the fact remains that even in these cases we are no nearer to the nuclear family. A dominant male helps in bringing up the young not in the role of daddy—since he has no stable or exclusive pairing relationship with any one female—but in his political and pedagogic role as one of the elders of the tribe. This type of society bears no relation at all to our own.

At some point, and for some reason, we diverged from the customs of our kith and kin and embarked on the path that led to Bide-a-wee and all the other separate little boxes. As to when it happened, the evidence points to the Pleistocene, the heyday of the hunting-gathering hominids, the era that finally produced true men. Why it happened is not nearly so clear.

The Tarzanists work from the logical premise that we must begin with something even simpler and more basic than the nuclear family of daddy-mummy-and-the-kids. So they take, as the irreducible minimum, the one-male, one-female pair, and try to account for its existence. Naturally, they put it all down to sex, and the needs of the male hunter. He had to make sure his mate would be faithful to him while he was off on the long trail, and all that.

But surely this is an extraordinarily subjective statement to make. Most primates don't give a rap whether their mates are 'faithful' to them or not. The most ferociously dominant males will often watch one of their favourite females copulating a few yards away without batting an eyelid. It's true the gibbon drives away all potential rivals, and we tend to assume that he can't bear them near his wife; but it may well be that he can't bear them near *him*, for he drives away his daughters as well as his sons. He just doesn't like anybody very much.

The truth is that there is nothing inherently exclusive, or inherently permanent, about the sexual relationship. There is no inbuilt reason why any creature should require a sexual relationship, any more than a back-scratching relationship, to build up into an enduring bilateral partnership. Of all the various bonds that enable animal societies to cohere, the sexual bond is by far the most likely to be ephemeral. Where it is not, there are always other, and more powerful, factors at work.

Let's take another look at that premise: that the nuclear family begins with the male-female pair. If you are thinking in terms of confetti and wedding bells, this is one of the truths we can surely hold to be self-evident. But if you are thinking in terms of social evolution, then so far from being self-evident it is blatant nonsense.

The irreducible minimum which preceded the nuclear family by many millions of years was not the daddy-and-mummy group at all. It was the mummy-and-the-kids group.

Michael Chance and Clifford Jolly: 'The mother/infant relationship is a universal relationship throughout the primates, constituting a stem structure in the societies of sub-human primates surpassing in cohesiveness even that of the male cohorts.'

The strength and importance of this stem structure was recently described by S. I. Washburn as 'a major discovery of the last few years'. It is not a question of the milk supply, nor even of the body contact which psychologists have now established to be also vital to healthy development. The need of milk and contact comes to an end at an early stage, but the relationship endures.

In rhesus monkeys, for example, it is retained as a powerful bond well into the adult life of the offspring. Donald S. Sade observed one case where 'a mother protected her offspring by diverting an adult male from attacking her offspring, when the offspring himself was fully mature'.

Moreover, the bond survives the birth of siblings, and as we move up the primate scale its strength increases rather than diminishes. The writings and films of Jane Goodall and her husband depict the life of chimpanzees in the wild, and Irven de Vore's comment on them was: 'What strikes me in these—especially in the recent films—is the extraordinary degree to which the offspring continue to be attached to the mother well into young adult life. What one finds is not a band or organized group in any monkey sense, but a great many older females surrounded by immature offspring—up to as many as four, including young adult males.'

Studies of monkeys in Japan and rhesus macaques in India establish that in these species also a similarly powerful bond exists.

So we now know that in chimpanzees, for example, an interaction once described as a mother-'infant' bond, and believed to be largely onesided and mainly nutritional, is in fact a strong reciprocal relationship which can last for upward of eight years—a long stretch in any primate's life. The more sentimental Victorians would have found this rather sweet; in our present period of slightly panicky backlash against mother worship, we may find it somewhat embarrassing. But the fact remains that in their natural state, and with no Freudians to point out that they are sick sick sick, this is how these uninhibited apes actually behave.

I find it impossible to contemplate this powerful and universal primate structure, which in the more advanced species holds together in an enduring blood relationship a parent and one, two, three, or four offspring, and lasts until those offspring are fully adult, without being driven to the conclusion that here is the origin of the human nuclear family. All that is missing is Daddy.

While primate family organization remained matrifocal, the males played an important role as leaders and defenders of the group and its territory; they were important to the clusters of young growing males as mentors and models; and their sex life was promiscuous and satisfying. It would never have entered the head of any of these free and splendid creatures to attach himself permanently to one individual female and the offspring she regularly produced and surrounded herself with. Yet ultimately that is just what he did: we must ask ourselves why.

There are many species in which the father plays a more than genetic role, and as so often we cannot draw a clean black line around the factors which dictate this type of behaviour. Some determinants, however, are easy to perceive.

They are all species in which the young are unable to fend for themselves, like babies, or nestlings, or wolf cubs. But this is not enough; for baby monkeys and baby kangaroos are helpless, too, but they arouse no paternal feelings.

They are all species in which the young are nurtured in one place—some nest, lair, or rocky ledge to which the parent must return. This appears to be a vital factor. I can think of no exceptions to it. Indeed, Robert Ardrey makes out quite a cogent case for believing that attachment to place is the really basic instinct in these cases, and devotion to the female (and sometimes the young) temporarily inhabiting it is merely a corollary.

Yet even this is not enough, for kittens and bears and rodents are also helpless in hideaways, yet Tom Cat and Father Bear and Bre'r Rabbit cling to their carefree bachelor ways.

It would seem that Daddy is brought into the picture only as a last resort, when for some reason or other the job of caring for the young is too exacting for one parent to cope with alone. This applies to many birds, where the young have to do all their growing in one short season and have appetites so voracious that it takes both parents all the hours God sends to fill their gaping beaks. It applies to the beaver, for hearth and home and the safety of the nursery depend on his twenty-four-hour repair service for plugging any breach in his dam at five minutes' notice. It applies to some Arctic species that live in an environment so ferociously hostile that without a shift system a parent could never hatch a chick, let alone rear it, without dying in the attempt of exposure and starvation. In circumstances such as these, Father rises to the occasion magnificently: he is a tower of strength. Where the living is easy he is apt to copulate and then go whistling on his way—anything that happens after that is strictly the female's affair.

Of all these factors, how many applied to early man? His young were born helpless, but only marginally more helpless than a gorilla's. However, as their brains grew more complex and they needed larger skulls, they had to be born at an increasingly 'premature' stage or they'd never have passed through the pelvic ring at all; so their period of helplessness was much longer, and since their mother had no fur for them to cling to they were more hampering to carry around.

Secondly, the lair. As the hunting-gathering economy became the norm, both males and females often found themselves in possession of more food than they needed to eat at that time. There had to be somewhere to take it to and store it; and at a later stage somewhere to cook it. It was already, in subhuman primates, a part of the pattern of the matrifocal family that the mother would share food with her young: this happens with chimpanzees. On the rare occasion when a chimpanzee captures himself some meat he will share it with his neighbours; but a mother will share food of any kind with her offspring if they request it.

Thus as the female hominid became more efficient at collecting food and more ingenious at processing it, the young would get to know that if they were hungry and nothing tasty was in sight they could always go to Mum for a handout. They would go to the place where she kept her containers and her grinding stone: they would go, in fact, home. And if they'd found something exciting but not immediately edible, like a tortoise, they'd take it home with them and she'd deal with it.

Remember, this was a relationship that even in apes lasts for eight years or longer. Any hominid who's spent the first decade of his life going home to a woman for his meals would have got into the habit of it. He'd begin to get the idea that that was one of the things females were for, and when his mother died or the matrifocal bond finally weakened he would automatically look around for another female.

Naturally he was not as dependent as the children. Very often he went off and caught small animals and ate them in the bush, or large animals and brought them home, the way he used to bring small rodents and tortoises. But when his quarry outran him or his spear missed or the spirit failed to move him, he liked to think there was somewhere he could go for a snack.

He got terribly annoyed if he found any adult males hanging around the same female and helping themselves.

Sex was one thing—there was plenty of that around and his own supply was not diminished if someone else had been there first—but food was something else again. The proceeds of two or three hours' gathering and grinding could be demolished in ten minutes by a hungry interloper, and when it was gone it was gone. Before long he found it wiser to take a stand on the sex issue, too; because since sex had got less sexish and had had love dragged into it, the woman tended to look with a kindly and quasi-maternal eye on anyone who cuddled up and treated her tenderly, and would allow them to dig into the melon-seed cookies without demur.

This infuriated him. Sometimes he devoted so much attention to preventing it and asserting his own prior right of sexual relations with her that he had little or no time left to go around being promiscuous on his own account. 'This is *my* woman,' he told himself. 'And this is *my* place. And these pots the woman has made are *my* pots. And these babies the woman has made,' though he had no idea that he had helped in their creation, 'these are *my* children.' He had horned in good and proper on the matrifocal family group, and was well on his way to becoming Daddy.

But true man, homo sapiens proper, is the child of the Pleistocene. In those turbulent millennia when the hominid wandered over the face of the earth, when the northern hemisphere pendulumed between ages of ice and ages of greenness, and the southern hemisphere between dusty dearth and torrential rains, then the third factor came into play. If the nuclear family had not acquired a father by that time, it would have been necessary to invent him. Again and again the human family must have passed through eras of climatic crises when almost every living child had a father whose devotion to the family group was unshakable—not because there is some reserve of nobility in the human heart which adversity invariably calls forth, but simply because a baby with any other sort of father would have died.

The era of Man the Hunter was undeniably a crucial one, but I believe it has been misunderstood in many ways. Above all, the length of it has been overestimated —men have tended to imagine that their ancestors were learning to hunt on the African plains throughout the missing ten million years when Hardy maintains they were elsewhere. If we accept the aquatic theory, then the time between the return to the land and the switch in most human communities to agriculture and stock-breeding is telescoped into a couple of million years. The only recognizable physical trace it has left on us is the ability of the male to throw things harder and more deftly than the female can. It makes them very good at cricket and baseball.

I believe, with the Tarzanists, that it was the era when the nuclear family began to evolve. But I believe its evolution had more to do with economics than with sex; I believe it has had a much shorter time to get under way than is often assumed, and I believe that in biological terms we are very imperfectly adapted to it.

If we decide that marriage is a system worth preserving —and there is something to be said for this view, although currently the iconoclasts are often more eloquent than the traditionalists—it will be no help at all to assume that in this endeavour Mother Nature is in there pitching for us at some subliminal level, and that all she needs is a little more help in the bedroom athletics department to make everything plain sailing. She's pitching for the gibbon and the beaver and the painted shrimp, but not for us.

Perhaps there is one more adjustment we have to make before the monogamous system can be made to work smoothly. For there is one more peculiarity of permanently pair-bonded mammals that we have not yet commented on. It is admittedly not true of pair-bonding birds, and pair-bonding mammals provide so small a sample that we cannot draw any firm conclusion.

It is interesting, all the same, that in several of these species the normal dominance-relationship of male and female is utterly absent. Robert Ardrey in his book *The Social Contract* mentions that 'the gibbon is the only [primate] species we know in which the female even nears the dominance of her mate'. The phrase 'even nears' was obviously as far as he could bring himself to go; but he derived this information from the studies of C. R. Carpenter, and what Carpenter actually wrote was: 'It may be concluded that in this primate, though the adults are very aggressive, there is an equivalence of dominance in the sexes.'

In the case of beavers the females are the more dominant and the more strongly territorial. Lars Wilsson says: 'When the female wants to attract a male, she first deposits castoreum inside her future territory, which gives her the stimulus she needs to get quickly the necessary domination over her intended partner.' When breeding beavers he found it best to pair a large female with a small male, as this settled the issue more quickly; but regardless of this the outcome was a foregone conclusion as long as the female was able to mark out her territory before the encounter. 'When Findus came rushing at her she calmly stood her ground and gave him the reception he deserved. After a few violent wrestling matches Findus had to admit defeat by the considerably smaller female, but she had to fight hard for several nights before she could feel she had the fat, spoiled Findus properly in his place....'

It seems a pretty terrifying nuptial arrangement, but for beavers it is the prelude to living happily ever after. 'Once the often somewhat violent pairing ceremonies are over, there is never any trace of discord to be seen between the partners. They sleep curled up close together during the daytime, and at night they seek each other out at regular intervals to groom one another, or just simply to sit close side by side and "talk" for a little while in special contact sounds, the tones and nuances of which seem to a human expressive of nothing but intimacy and affection.'

The union lasts as long as they both live; a male who loses his mate shows signs of intense grief: 'He ate nothing for more than a week and simply roamed restlessly up and down in his enclosure'—and in the wild state, unless firmly taken in hand by a strapping widow, he is apt to end up marrying his eldest daughter and settling down to a further stretch of domestic bliss. Apart from this touch of incest the married life of the beaver is exemplary.

The *callicebus* is another example of a monogamous pair-bonding animal, and in this case also there is no sexual dimorphism, and the females are no less likely to be dominant than the males.

It may of course be pure coincidence that monogamy in the world of mammals it is so often found to coincide with as it were, some degree of women's liberation. It may be that the habit of monogamy has eroded any male dominance that once existed. It may be that the habit of male dominance evolved chiefly to regulate interactions between males, and a male who dwells permanently in the bosom of his family has no need of it, and finally discards it. Or then again it may be that a measure of sexual equality came first, and is a necessary prelude to really successful pair-bonding in a mammal species.

If this is the case we have a long way to go, because the human male's preoccupation with dominance has survived all his evolutionary vicissitudes on sea and land. It is as strong in him today as it was in his ancestors when they howled in the treetops, and it will take us a chapter just to scratch the surface of it.

Nine

Primate Politics

We are now entering the realm of primate politics. This, like sex, is an area where the protagonists are apt to lose their cool. Purely abstract concepts, such as the nature of aggression, and heredity versus environment, are likely to be debated with an air of tightly leashed politeness, and with suppressed mutterings of 'Anarchist!' 'Fascist!' 'Marxist!' clearly audible between the lines. I am as little likely as any of them to attain a godlike objectivity, but I assume this is the kind of rumpus where anybody can join in—and anyway 'Feminist!' will make a refreshingly new epithet for ethologists to hiss.

When you begin reading about primate social behaviour and its relevance to human evolution, the first thing you will notice is that everybody seems to be talking about baboons and macaques—especially about baboons.

At first this seems a little odd, because biologically we are not very closely related to these animals. They aren't even apes—they are only monkeys. Yet on these two species exclusively most popular discussion of human social inheritance has centred. The index to Robert Ardrey's *Social Contract* gives thirty-eight lines of references to baboons and macaques, whereas no other primate—ape, monkey, or prosimian—has more than four. Lionel Tiger's chapter in *Men in Groups* on primate male bonding concentrates wholly on these species except for a brief discussion on langurs. He spends no time at all on the apes.

Now, why should this be? Admittedly the baboon is a popular and successful species, and since he is largely a ground-dweller he is easier to study than some others. But

chimpanzees and gorillas, being much more closely related to homo sapiens, have attracted the concentrated
attention of many observers recently, and we have learned
a great deal about their behaviour. It is puzzling that the
popularizers have so little to say about our nearest kith and
kin.

It seems to me they have taken a look at our kith and
kin and rapidly concluded that the way chimps and
gorillas behave doesn't *explain* anything. It all depends,
you see, on what you are setting out to explain. If you
are starting out with the premise that man is the most
aggressive and bloodthirsty creature on the face of the
earth, then these cousins of ours will be nothing but an
embarrassment to you. A few quotations will be enough
to make this plain.

First, the gorilla. He is, says Ardrey, a 'gentle, inoffensive,
submissive creature for whom a minimum of tyranny
yields a maximum of results'. Irven de Vore calls him a
'mild-mannered vegetarian who likes to mind his own
business' and lives 'in a state of mild and amiable
serenity.... His dominance over the group is absolute, but
normally genial.... The leaders are usually quite approach-
able. Females nestle against them and infants crawl happily
over their huge bodies. Amity reigns. When a band of
gorillas is at rest, the young play, the mothers tend their
infants, and the other adults lie at peace and soak up
the sun.' You can see that's not the kind of stuff that
sells newspapers.

And the chimpanzee is just as frustating for the blood-
and-thunder boys. Ardrey: 'The amiable chimpanzee
seems to found his society on nothing very much but his
own good nature. There is an order of dominance but it
is not at all severe. When band meets band in the forest
or on the savannah, there is enormous excitement but
no antagonism, and all may wind up feeding in the same
trees. The chimpanzee has demonstrated, I presume, that

we must reckon on some degree of innate amity in the primate potential.... The chimp is the only primate who has achieved that arcadian existence of primal innocence which we once believed was the paradise that man had somehow left....'

De Vore: 'Chimpanzees are the most compliant of the apes. They revel in applause; they love attention.... They learn to control their emotions. In the wild, a young chimpanzee learns as it matures not to irritate the adults. As a juvenile, it learns to control its natural exuberance in play with infants of the group so as to avoid injuring them....'

These two, I repeat, are our closest evolutionary relatives. You might reasonably expect to be able to go into a library and find three or four books explaining that this is why *homo sapiens*, by and large, is also a mild-mannered and unassuming species. You should be so lucky!

No, no—the baboon is the one you are invited to contemplate. Ardrey hurries the chimp offstage with a parting sneer to the effect that it's all very well to be amiable but see where it's leading him, he's an 'evolutionary failure' (as though only amiable creatures became extinct). Then he settles down to the interesting part:

'The student of man may find the baboon the most instructive of species. Among primates his aggressiveness is second [sic] only to man's. He is a born bully, a born criminal, a born candidate for the hangman's noose. He is submissive as a truck, as inoffensive as a bulldozer, as gentle as a power-driven lawn-mower. He has predatory inclinations and enjoys nothing better than killing and devouring the newborn fawns of the delicate gazelle. And he will steal anything....' And so on. While his male reader avidly polishes his spectacles and thinks: 'Yeah, that's me all right. Tell me more about the bulldozer and how I ravaged that delicate gazelle.'

He reads on and learns how the male baboon is twice

the size of the female. He keeps a herd of them in terrified subjection; he is fiercely jealous of them when they are in oestrus; if one of them strays he will punish her severely and fight any intruding male; if he is strong enough he will hog all the best food and impose his will brutally on weaker males. He demands instant and unquestioning obedience, and when danger threatens he will marshal his troops and stand up and fight like a hero, shoulder to shoulder with his loyal comrades.

He is not really very much like the man who is reading the book. But the man who is reading the book (to say nothing of the man who is writing it) gets no end of a kick out of thinking that all that power and passion and brutal virility is seething within him, just below the skin, only barely held in leash by the conscious control of his intellect. He used to like reading about gorillas when we judged them by their faces and their roaring, but the more he learns about them the more he begins to suspect they're a little bit wishy-washy; so he averts his eyes from the primate family tree, forgets that he descended from apes, and identifies with the baboon even if it means making a monkey of himself.

A few general facts about the structure of anthropoid societies will help to put the picture into perspective. We will omit the very rare species, like the gibbon, whose society consists only of the nuclear family, and concern ourselves with the vast majority which form into larger bands or troops.

They fall into two major categories—according to whether their societies are acentric or centripetal. I have taken these terms from a detailed study of primate societies by Michael Chance and Clifford Jolly. An acentric society is individualistic and loosely structured. (If you favour it you can call it democratic; if you disapprove you can call it anarchic.) A centripetal society is highly structured and organized around one or more alpha-male leaders. (If you favour it you can call it a society of law and order; if you disapprove you can call it tyranny.)

A favourite example of the acentric type is the patas monkey, and a favourite example of the centripetal is the baboon.

The factor that splits them right down the middle is how they react to danger. If you are a patas monkey and you venture out into the open you remain all the while keenly alert. Even while you advance, your eyes are darting around, checking on everything in your environment, making sure the way is open for you to retreat the way you came, searching for an even quicker way up into the branches if one should be available. At the swish of a tail or the sound of voices you can be off like a shot, and all your equally alert companions will do the same. The whole band of you will scatter like the sparks of an exploding firework and you'll all end up high in the branches, and safe.

But if you are a baboon you will take a diametrically opposite view. You pay comparatively little attention to your environment. You may be in the middle of a plain where the environment is pretty featureless. You may feel that it's a long way between trees, and a hungry leopard or an angry farmer could easily bring you down in your tracks before you reached one. If you scatter you only make it easier for a predator to pick you off. Much the best bet is for everyone to stick together, and the safest place to retreat to is the vicinity of whichever of your comrades has the sharpest teeth and the most courage. So when *your* eyes are darting, what they are checking up on is where you are in relation to the alpha male or males. If he moves on, you can't afford to be left behind; and if he signals some danger you haven't been aware of, you'd better believe he knows best, and close ranks, and do what you're told.

Clearly the position of males in these two societies will

differ radically. If a patas troop scents danger the first thing
that happens is that the single male gets as far away as
possible from the female assembly and while they scatter
he puts up a display of bouncing around to divert the
attacker's attention before running away himself. They
don't have to obey him—they simply have to disperse.
Scattering is their answer to everything. Patas monkeys
have no submission gestures because if they are threatened
they simply run; but they are pretty non-aggressive anyway, and the high-level threats of baboon society are unknown.

Roughly speaking, this is the sort of behaviour displayed by the 'bandarlog', the monkeys who drove Rud-
yard Kipling wild with irritation and contempt, because
Kipling was a pukka sahib and the bandarlog were an un-
disciplined mob who couldn't even co-operate or concentrate, or (as he portrayed them) even finish a sentence:

> Now we are going to—never mind!
> Brother, thy tail hangs down behind!

Among baboons, on the other hand, the males have to
be bullies. The troop has to rally to them. And since not
even baboons are *born* disciplined, discipline has to be
inculcated. Females, juveniles, and subordinates have to
be taught their place, and frequently reminded of their
place, by threats and punishments and bites on the neck.
Usually they learn fast, and a display of fangs or the
flash of an alpha's eyelids is enough to keep everyone in
line. As long as they remain in line the alpha's rule is
benevolent and administered with rough justice and old-
fashioned chivalry. Mothers with infants are always flanked
by protecting male outriders when the troop is on the
move, and in disputes between subordinates the leader
sides with the weaker of the two. (Perhaps chivalry isn't
quite the word—if the dispute is between a female and a
subordinate male he'll usually back the male.)

If we can learn nothing else from baboon society, we can

at least learn to discount the idea that the human capacity to co-operate and form a disciplined, highly structured society can only have evolved because homo sapiens had to learn to 'co-operate on the hunt'. Such structuring is clearly present among baboons. Deployment of a baboon troop on the move is governed by a complex system of rules dictating where everyone is in relation to the central hierarchy according to age, sex, status, competence, and other variables. And yet the baboon is from 90 to 98 per cent vegetarian, and such small-scale slaughter as he does occasionally indulge in is never 'co-operative'. The urge to relate man's co-operative potential to his diet is simply another Tarzanist aberration.

Lionel Tiger has a theory that baboon-type political systems are the consequence of 'savannah as opposed to arboreal life'. This is a very nice theory for anyone who wishes to equate human with baboon society, because it suggests that though *biologically* we are like chimpanzees, we became socially more and more baboonlike as we left the jungle behind us and moved on to the plains.

We'll return to this theory later, because I believe there may be something in it; though there are quite a number of facts that don't fit in. For example, the archetypal acentric, the patas monkey, is himself a ground-dweller. He frequently inhabits precisely the same territory as the baboons, and indeed penetrates northwards into areas more arid than the baboon can cope with. On the other hand, there are centripetal species which never leave the trees if they can help it.

The next big question, since homo sapiens relates most closely to the apes, is: how do we classify the gorilla and the chimpanzee? Are they acentric like the patas, or centripetal like the baboon? In fact they don't behave very much like either; but two of the criteria for identifying the basis of their social structure are (a) is there a leading male, or signs of a ranking order among males?; and (b) when there is danger, do they scatter or do they clump together?

The answer on both counts suggests that, like the baboons, apes form centripetal societies. The signs of this are not so clear and unequivocal as among the baboons—except for the very obvious dominance of the alpha gorilla—but if you keep a close watch on the attention structure you will be left in no doubt; they are centripetal.

We are now confronted with two major questions: Are our own social instincts analogous to theirs, and how do they manage to make a centripetal society work without all the slashing and snarling and bullying and cowering that is such a feature of baboon social interactions?

The difficulty in comparing human society with any other animal's is that the cultural components in it are so powerful that they tend to blur any inherited instincts that may exist. If you are examining an English public school, or a Nazi S.S. troop, each of them in its different style would convince you that our instincts are those of the baboon. If you consider the aggregations of human beings at Woodstock or the hippie colony at Haight Ashbury, or the reactions of people in an earthquake, you would feel tolerably certain that our instincts are those of a comparatively amorphous troop, like the patas.

Fortunately, however, it is quite easy to find social groups of human beings which behave in exactly the same way all over the world and in every type of culture. I am referring, of course, to our juvenile clusters—unsupervised groups of children up to the age of about six.

Adriaan Kortland gives a vivid account of chimpanzee response to a potential threat, which he elicited by placing a stuffed leopard in a strategic situation and observing the chimps' reactions. If you have ever watched (or in your childhood participated in) a group of small children reacting to an unfamiliar, slightly alarming animal—say a large grass snake—you will be struck by the resemblance.

'Following a moment of dead silence on catching sight of the leopard, there was a burst of yelling and barking, accompanied by every member of the group charging about in different directions. A few fled, but returned soon afterwards to join the majority, who began leaping up and down and charging the leopard, brandishing big sticks or broken-off trees.... Some of the blood-curdling barking was loud enough to wake a human neighbour 600 yards away.... Interspersed with these communal or individual charges were periods of seeking and giving reassurance by holding out the hands, touching their neighbours.... Voiding of diarrhoea and enormous amounts of intense body scratching took place. The attacks on the leopard were more or less rhythmical, and followed by brief increases in fear symptoms and the seeking of reassurance and longer periods of sitting down watching the leopard. The aggressive aspects gradually waned after an hour, being replaced by intense inquisitiveness.... One chimpanzee poked it with its fist, another smelled it, and finally the leopard's head was detached from the body and rolled about. Another chimpanzee seized the tail and they all rushed off into the bush with the body.'

This is not at all like the baboon, and not at all like the patas monkey, but it seems to me to be more like unprocessed humanity than either of them.

As to how the apes' centripetal society works, the essential key to being high in a primate ranking order is the ability to command the attention of other members of the troop. There are two ways of doing this, as has emerged from the work of H. B. Virgo and M. J. Waterhouse, and V. Reynolds and G. Luscombe.

The two methods are classified by Michael Chance as the agonic mode and the hedonic mode. The baboon aspiring to dominance commands attention by biting and threat displays with his huge canine teeth. This is the agonic mode, and is a pretty reliable one. If someone slashed your lip open yesterday you are at least going to be acutely aware of him next time you pass him in the street, especially if he pulls his knife again and snarls.

Robert Ardrey is rapt with admiration for the primate society constructed on this principle and uses it for pointing a moral to misguided liberals: 'It is as if, buried in the baboon subconscious, the truth, like one's shadow, can never be far away.... The secret of his success would lie in that undistinguished, unwashable brain. The baboon will never persuade himself that aggressiveness is a product of frustration. The young will never blame their failures on lack of parental love in infancy. Should the proposition that competition is somehow wrong come baboon way, small brains would be dumbfounded; should some mutant baboon idealist insist upon it, he would be greeted with the lifted eyebrows not of human surprise but of monkey threat....'

That's one in the eye for permissiveness. There is, however, a drawback to the baboon's agonic organization. The brain is undistinguished and unwashable, all right; no subversive heresies will ever penetrate it. But it is so totally unwashable that it is unlikely ever to become any more distinguished. The structure is so rigid that true communication is at a minimum—it is limited to the ritual posturings of swaggering arrogance on one side and cringing submission on the other.

The typical encounter between superior and inferior is brief and leads to flight, or withdrawal to a respectful distance. This immediately ends the interaction. The whole complex system depends on everyone's performing his or her stereotyped role, and any new behaviour pattern which might disturb this, however adaptive it might prove if generally followed, would never get off the ground. What Ardrey calls the baboon's 'thunderous evolutionary success' has been achieved by the same means, and at the same cost, as the thunderous evolutionary success of the termite colony. They do their own thing to perfection, but they have left no options open. It is unlikely that they will ever go on to do anything else.

The hedonic mode favoured by the apes is quite different. Here again, a high place in the ranking order is attained by an outstanding ability to command the attention of one's fellows. But the apes are more advanced than the monkeys, and they have made a discovery which perhaps more than anything else made possible our own dramatic mental forward leap. They learned that you don't necessarily have to bite someone in order to make him take notice of you. Among gorillas and chimpanzees this type of physical aggression is extremely rare.

So how do they do it? Primatologists call it 'display': to put it in the simplest possible terms, they do it by showing off. They seek for ways of making themselves conspicuous: they bounce around and shake the branches. They find interesting objects, and their companions cluster around to see what they've got and what they're going to do with it. The dominant gorilla, the alpha silverback, of whom much is expected, has been seen to mount the most stupendous show-stopping performances in this line.

First he starts by hooting. He gives anything up to forty hoots at an increasing tempo. He picks a leaf and puts it in his mouth. He stands up on his hind legs, grabbing a handful of vegetation and throwing it into the air. Then he beats his chest up to twenty times with his hands, using them alternately, and slightly cupped. Sometimes for good measure he kicks one leg into the air while he is doing it. Immediately after the chest beating he starts a curious sideways run, first a few steps bipedally and then charging sideways like a gigantic crab, sweeping one arm through the vegetation, slapping at the undergrowth, shaking branches, breaking off or tearing up whole trees. Finally he thumps the ground, usually with one palm but sometimes with two, as who should say 'Follow that, buster!' Of course no one can, though even infant gorillas of six months old have been known to rise shakily on to their hind legs and slap at their puny chests, watched by their mothers as fondly as if they were aspiring Hollywood tots at their first audition.

You can see it's no accident that the chimp in captivity willingly learns new tricks and loves applause. You could waste a lifetime trying to teach a baboon to ride a bicycle, or even perform some simpler trick more within his capacity. He just doesn't see what's in it for him. But for the chimpanzee it is his method of acquiring dominance. To extend his repertoire improves his ability to attract attention, and as long as fascinated eyes are focused on his efforts, even though they are human eyes, he has a rewarding belief that he has added to his status. And he's right, isn't he?

Two major advantages accrue to the hedonic mode. Firstly, while dominance by threat stultifies social interaction, dominance by display promotes it. Threats cut individuals off from one another, but display brings them closer, to watch and investigate and congratulate. As Michael Chance comments: 'Display behaviour, responded to by greeting, stimulates and enhances the tendency of individuals to develop many forms of contact behaviour or behaviour at close quarters. Manipulation, not only by grooming, but also by holding and investigation, is jointly engaged in.... Their attention may also switch to the environment or to other objects and give rise to manipulation of objects as tools.... In the hedonic mode, display leads to outgoing but flexible social relations which can act as the medium for the dissemination of information within the society.'

The second major advantage is the incentive given to behavioural innovations which may prove advantageous to the species. The baboon's neck-biting is a good trick as far as it goes, but a bite is a bite is a bite; whereas the young ape, competing for attention with other young apes, is daily stimulated to search for something new by the troop's constant unspoken challenge: *'Etonne-moi!'*

Which mode was the hominid's society cast in? It seems glaringly obvious that only the hedonic mode could have led us to where we are today. But if you want confirmation, I urge you again to consult the largest available pool of virtually uninhibited human social interaction. Most males, through no fault of their own, have not had the ethological advantage of spending up to ten years of their lives in the constant company of the young of their own species, but they are still free to stroll along to the nearest nursery school or infants' playground and observe what happens when children first begin to construct a social framework for themselves. For every encounter which takes the form of two small boys hitting each other, there will be fifteen or twenty which are causing the whole yard to ring with cries of 'Look at me!' 'Hey, watch this!' 'Look—can you do this?' 'Look at me, I'm a cowboy!' 'Look what I've done!' 'Look what I've found!' 'Look at my new doll!' 'Come and see Johnnie, he can stand on his head!'—while the unstable Johnnie is gasping 'Everybody look at me—quick!'

You don't have to be outstanding in size or courage or aggression to make your mark in such a society. If you're double-jointed, or can wiggle your ears, or draw better pictures, or turn better cartwheels than anyone else, then you've got status. It doesn't matter what it is as long as it ensures that when you raise the cry of 'Look!' somebody will look.

So, with all due respect to the baboon and his admirers, I submit that *homo sapiens* as a social being is modelled ineradicably on the hedonic mode of dominance by display, and that basically our relations with our fellows resemble more than anything else those of the chimpanzee, with all that implies of amiability, flexibility, curiosity, and exhibitionism, as well as the tendency to react to sudden peril with a slackening of the bowels and a desire to hold somebody's hand—or even sometimes dive for cover without waiting to warn the whole troop, a piece of turpitude no baboon would ever be guilty of.

I think the agonic mode was the more primitive one. Of course, it was not entirely eliminated and replaced either in apes or, certainly, in ourselves. In moments of rage, or frustration, or when frightened and cornered, or when dominance rivalries become too acute, or in interactions with another species, the agonic mode is still resorted to. It appears to be resorted to more readily by males than by females, partly because they are more concerned with dominance, and partly because as defenders of the tribe they would be more likely to confront external threat from other animals. It is clearly not very adaptive to show off to a leopard—though even in such encounters as these a spot of roaring and chest-beating and bouncing about will sometimes work wonders, and it has been known to scare the daylights out of the gorilla's only really dangerous enemy, *homo sapiens*.

This is why I agree with Anthony Storr in his contention that the space race, however costly, is to be welcomed between superpowers obsessed with anxieties about their relative rank order. Some day we must find a way of curing them of the obsession, but until that day comes we must count it as a step forward that while the H-bomb was an agonic signal, the moon landing was a hedonic one. Nation-states still behave far more irrationally than most of their individual members, but we may cherish a small hope that one day they may catch up with the chimpanzee.

So much for the basic structure of the social group. The next feature to examine is the subgroup: the male cohorts, the female assemblies, the juvenile clusters, and the matrifocal family groups. The juvenile clusters—bonds of co-evals who play together—have changed hardly at all; and the matrifocal family groups have evolved into nuclear-family units.

Lionel Tiger's fascinating and widely acclaimed book *Men in Groups*, examining the phenomenon he describes as 'male bonding', has lately stimulated discussion about the male cohorts, while the female assemblies have attracted no attention at all. As far as most people are concerned they cannot be said in any real sense even to exist, because 'females', Mr. Tiger declares, 'do not bond'.

The facts are as follows: in all primate societies (ex-cept the rare monogamous gibbon-type ones) there is a strong tendency for females to aggregate together. There is also a tendency for males to aggregate together; these are often, as with the savannah baboon and the gorilla, groups of breeding males with females and young ac-companying them.

However, of the two sexes, a primate male is far more likely than a primate female to opt out of the company of his own sex altogether, and place himself at the head of a harem of females and their young, as happens with the patas monkey and the hamadryas baboon. In this way one male will monopolize up to nine females, which upsets the sex ratio, so the surplus males aggregate together into a second type of male cohort called a 'bachelor band'.

Mr. Tiger has a lot of perceptive and penetrating things to say in the later sections of his book, but in the early part of it he makes a number of statements which aroused resentment in a lot of people (especially women) and which are in themselves greatly misleading. Let's get these out of the way first.

To begin with, he manages to give the impression that male bonding is a more or less ubiquitous phenomenon in animal societies. True, he does at one point make the rather disarming confession: 'I began this project thinking that male bonding among non-primates was more common than I found it in fact to be.' But he doesn't make it clear whether, when he searched for it among non-primates, he found it merely uncommon, or in fact *totally unknown*. Certainly he fails to come up with a single non-primate example. He merely urges field workers in general to try a bit harder, because it's sure to be there somewhere if they only look.

From there on he confines his discussion to male bonding among monkeys. And just what does he mean by the male bond? His definition reads: 'Male bonding among primates is defined here as a particular relationship between two or more males such that they react differently to members of their bonding unit as compared to individuals outside it.'

This is fine as far as it goes, which is not very far. It does not satisfy him for long, because this definition would apply, *mutatis mutandis*, to matrifocal family groups, female assemblies, juvenile clusters, or any other type of primate grouping. He wants to narrow it down, to exclude anything with females in it.

He does this by drawing a distinction between 'bonding' and 'aggregation' (as found among herds of ungulates). Aggregation doesn't count because 'there is no selection involved'. He illustrates his meaning with a human parallel: 'A male will not bond with just any member of his ethnic, religious, familial, or social-class group; he will bond with particular individuals because he has certain prejudices and standards in terms of which he is willing to bond.' He is talking about the Freemason syndrome.

In other words, he says the bond is selective. He quotes, as if at random, an example: 'In a baboon troop containing eight adult males, perhaps three will form a bond. Each of the three will be particularly sensitive to the other two, intensely aware of the distinction between bond and non-bond males, and committed to membership of the bond in so far as this leads to sociosexual advantages and responsibilities.'

Well, that cuts out the female primates, all right. The only trouble is, it cuts out 99 per cent of male primates as well; for there are only two kind of primates that display this cliquish behaviour. And you can guess who *they* are: our old friends the baboons and macaques.

Irven de Vore: 'The system of rule by clique or Establishment is peculiar to baboons and macaques, and one can easily see why it came into existence. Because the monkeys are so potentially aggressive, the peace in a large group can only be preserved by a force stronger than any one animal could command.'

What it comes down to is this: the only solid fact behind the 'sensitive and selective' theory of male bonding is that in the two most thuggish of monkey societies, baboons and macques—and *only* in these societies—the dominance structure is apt to be headed by a troika of dictators instead of by a single dictator. These alliances are not held together by personal regard, but by pure self-interest. Any slight shift in the balance of power proves them to be about as sincere and durable as the Russo-German Alliance in World War II. And our nearest relatives, the hedonic apes, don't form these cliques at all.

We are left with the fact that male primates form groups, just as female primates do; and that the behaviour within these groups differs. With this I entirely agree.

Differs how? Is it that in the male-male groups the relationship is closer, warmer, more companionable? Well, *pace* Mr. Tiger—no, it isn't. Any primatologist will tell you that for most of the time the reverse is more likely to be the case. One correlate of warm and friendly relationships between individual primates is the frequency of grooming. Females may groom males or vice versa; mothers may groom their infants; when the infants grow older they may return the compliment; a male may groom another male after a bit of an incident to confirm that friendly relations are restored; and so on. But in all species it is true to say that most of the grooming is initiated by females. And while they may enjoy grooming children or males—especially high-ranking ones—yet most frequently of all they groom one another.

Another correlate of friendly relations is spatial. Here again it is generally true to say that female assemblies are packed more closely together. Perhaps they trust one another, while each male warily preserves a 'personal space' around himself of more than an arm's length, to guard against an unheralded attack by one of his bosom friends. Or—since personal space is a sign of importance too, and everyone gives alpha males a respectfully wide berth—perhaps the females are, in anthropomorphic terms, merely less pompous.

Can the difference be, as Lionel Tiger at one point suggests, that the female groups are more 'emotionally labile', while the male cohorts preserve a stiff upper lip?

He supports this by quoting Michael Chance's argument that breeding males have to learn to inhibit their emotional responses while, says Tiger, it would be advantageous for females *not* to inhibit theirs, so that they might be 'uninhibitedly attuned to their young'. This is a strange line of argument. Michael Chance was talking about inhibiting aggressive emotions, not loving ones, and if the mother of a young infant didn't inhibit such emotions as impatience and irritation she would surely kill it.

The fact is that emotions in the female assemblies are in a relatively low key. In baboon societies, in agonistic encounters between male and female, the females admittedly screech rather a lot; if I were being pursued by a chastising male baboon I would screech too. But disputes between the females themselves are settled swiftly by bickering and scolding, possibly backed up by shoving and slapping.

In any case they are invariably over in a minute and forgotten. But a dispute between males, if the subordinate male does not quickly back down and pay homage, may lead to threat and counter-threat, eyeball-to-eyeball confrontation, and aggressive tension mounting gradually to a terrifying pitch and ending in violence.

Finally, around Chapter Seven, Tiger pinpoints the real, universal, and indisputable difference between cohorts of males and groups of females: namely, that the males are more aggressive. In a general sense this is true not only of primates, but of the males of the great majority of species.

And what is all this aggressiveness for? There is a popular belief that males are more aggressive and armed with formidable weapons in order that they may protect their loved ones from the terrible dangers that surround them. The male baboons will advance against the leopard and so on, while the mothers and children cower in the background. It is true, they will.

But although the weapons and the aggressiveness may be used for this purpose, let us not fool ourselves that this is primarily why they evolved. The stag has towering antlers and a powerful neck to withstand the shock of impact; the elephant seal has tusks that could rip you wide open. But for most of the year the stag wanders off and leaves the does and their young to fend for themselves, and once the jousting season is over, though predators may still be about, he lays aside his arms so that he can grow an even larger pair in time for the next tournament. As for the elephant seal's tusks, where he hauls out there is nothing and nobody he needs to use them against except other elephant seals.

Male weapons and male aggressiveness are for dominance, not for protection. Where weapons are designed for use against other species they are generally issued to both sexes indiscriminately, like the eagle's talons and the

snake's poison and the wolf's teeth. It is a pretty safe
rule among birds and mammals that the more impressive
the distinction between male and female, the less likely it
is that the male's physical superiority has any relevance
either to protection or to predation. The Indian elephant's
magnificent tusks are useful, zoologists tell us, 'for defence'.
Against what? His mate is smaller and has no such equipment, but it is hard to imagine any animal cheerfully pro-
posing on that account to have her for his dinner. Even
when the impressive difference is not in size but in sheer
belligerence, as with fighting cocks, you can be pretty sure
where the belligerence is going to be directed: against
other fighting cocks. The same is true of primates. And
of men. The aggressiveness is for fighting each other with.

That statement has an air of something like naïveté in
view of all the torrents of highly philosophical prose that
have poured from the presses in the last decade on the subject of aggression. Most of the writing about it has been
done by men; and it is a powerful temptation for them to
assume that anything that men have more of can't be all
bad. This has led to a subtle kind of 'don't knock aggression' campaign. It is true, of course, that aggression was
badly in need of a public-relations officer. Its stock went
down rather abruptly, especially among intellectuals,
when they counted up the millions of dead and maimed
after the late unpleasantness of 1939–45.

How do you do a rehabilitation job in a case like this?
It is largely a matter of definition. You begin by defining
the word so nebulously that it embraces almost any kind
of drive; you take the line that a child needs an adequate
supply of 'aggression' to motivate it to persevere and, for
instance, solve a quadratic equation or undo a knot in his
shoelace, where another child lacking in 'aggression' might
give up the attempt.

Here is a little prose poem by Robert Ardrey on the
subject:

'This is the aggressiveness that many would deny. It is the inborn force that stimulates the hickory tree, searching for the sun, to rise above its fellows. It is the inborn force that presses the rosebush to provide us with blossoms. It is the force, brooking no contradiction, directing the elephant calf to grow up, the baby starfish to grow out, the infant mamba to grow long.... We seek the sun. We pursue the wind. We attain the mountaintop and there, dusted with stars, we say to ourselves, Now I know why I was born.... Or we achieve a transcendent vision of heaven and earth and God. We find a scarred desk highpiled with old books and, enraptured, we discover in the musty past our shining selves. All is aggression.'

Now, I don't say this is nonsense. It is a beautiful inspirational rhapsody, for those who like that sort of thing, in praise of—what? Of Life, I suppose. And I'll drink to that, any time. I would only say that when I use the word 'aggression', and when most people use the word 'aggression', and when Ardrey himself uses it to say that male baboons are more aggressive than female baboons, the word doesn't really have much connection with rose blossoms and stardust and old books and transcendent visions.

Lionel Tiger has a rather more realistic stab at it: 'I define aggression as a process of more or less conscious coercion against the will of any individual or group of animals or men by any individual or group of people.' Coercion against the will of ... Well, yes. Up to a point.

But this definition would mean that if a man grabs the hand of his two-year-old son who is aiming to toddle over the edge of a cliff, the man is behaving aggressively. It would also mean that if he arrives at his house late and drunk and kicks his front door to splinters because he can't make the key fit, then he is not behaving aggressively. I find both of these conclusions very strange.

I would like to take aggression, as Masters and Johnson

took sex, out of the poetic realms where it is described as a many-splendoured thing, and out of the moralistic realms where what matters is who is doing what to whom. I would like to see it treated scientifically.

Aggression is that physiological phenomenon which in a mammal produces the following symptoms: secretion of adrenalin into the bloodstream, faster heartbeat, increased blood pressure, changes in the circulatory system resulting in less blood being supplied to the body surface and more to the muscles and brain, more rapid production of blood corpuscles, quicker coagulation of the blood, staring, quicker and deeper breathing, inhibition of salivation, of secretion of gastric juices, and of peristalsis, rise in blood sugar level, sweating, and erection of the body hair. In the well-known mammal *homo sapiens* all these symptoms still occur, although the last has become vestigial.

The whole thing is an admirable piece of mechanism preparing the subject for physical combat. When in this state of arousal he can move faster and strike harder, and if wounded he will lose a little less blood. In contexts other than physical combat it is far less serviceable, and often counter-productive. In other words, aggression is for fighting people with.

For instance, the child with his shoelace or his quadratic equation is not more likely but less likely to solve his problem if his heart is pounding, his breath coming fast, and his bloodstream being pumped full of sugar and adrenalin; and though the man who kicked his front door in attained his objective of entry, it was not by the best or quickest method. Physical processes may be toned up during a period of aggressive arousal, but the process of reasoning is impaired.

Admittedly, other emotional responses may have the same effect. Fear is another emotion, once a valuable life preserver, which in a civilized context often leads to mindless and maladaptive behaviour.

But there is one important difference between these two reaction systems. Nobody writes paeans in praise of fear. This is because nobody enjoys experiencing it. Whether at its lower chronic level of anxiety or its high-level intensity of panic, people find it distressing. They seek to dispel it in their friends. 'Don't be afraid,' they say. 'There's nothing to worry about.'

Of aggression the opposite is true. As a popular form of arousal it rivals sex. It invigorates; it gives a sense of well-being and increased stature; it is immediately emotionally rewarding. It can be secondarily rewarding, too, because people who behave aggressively tend to get more of their own way than people who don't. For these reasons there is a strong tendency for people (both male and female) to seek out and repeat situations that arouse feelings of aggression in them, or to return to them mentally and rehearse them over and over in their minds so that the delicious shot of adrenalin will once more flow through their veins.

To put it in the simplest terms, aggression can be addictive. We have no need to visit a chemist or use a hypodermic in order to inject into our bloodstream a dose that can blow the mind. We have a do-it-yourself kit. And in dealing with our friends we do not seek to dampen down this tendency as automatically as we seek to dampen down fear. We more often feel it as an act of empathy to stoke it up: 'Yes, I don't blame you ... it's outrageous ... I don't know why you put up with it ... He ought to be shot...' Only after a man has had his first coronary will his doctor suddenly put this stimulant on the danger-ous-drugs list, saying: 'Cut out the whisky and don't let yourself get worked up.' (The patient may be able to get help with his drink problem, but nobody's yet founded an Aggressives Anonymous.)

This is the stuff of which the male bond is composed. Lionel Tiger: 'I suggest that males bond in terms of either a pre-existent object of aggression, or a concocted one....'

'Male bonding is a function of aggression.'

As soon as an outside enemy appears, hostile encounters between members of the male cohort cease. All aggression is directed outward, against the enemy. And the rewarding sensation of hostile arousal against the enemy is deepened and enriched by the warmer and even more rewarding sensation of love for, and solidarity with, the brother in arms.

Here is a description by Dr. F. Kahn of baboons: 'Social attraction ... is fluctuating. The flock of monkeys lacks any cohesion whatsoever when material living conditions are good; aside from family relationships ... the individual ignores the community. But if a cracking of branches announces the approach of a leopard ... everything changes as it does for us the day a war is declared. ... A troop of monkeys in a state of war is something respectable.'

This is the origin of the male bond. From the inside it clearly feels wonderful. We know it must be rewarding (as we know vaginal orgasm in animals must be rewarding) from the eagerness with which males present themselves in places and situations where they expect this bonded feeling to be induced in them—cup finals, conventions of secret or semi-secret brotherhoods, mass rallies, and (at least, in the past) battlefields. We know it must be deeply emotional, because from the earliest cave drawings and the earliest sagas it is clear that war moved men to produce art and poetry much sooner and more prolifically than the love of woman. We know it must be a kind of love, and of a high order, because it produces deeds of devotion and self-sacrifice, and trust and obedience so unqualified that even the power of independent thought is sacrificed on its altar. 'Theirs not to reason why, theirs but to do and die.'

I am prepared to believe it is something I have never experienced and never will experience. I know that when I see a film of men in battle—and like most mothers of sons I have learned through this medium more about battles than I ever wanted to know—the sight arouses nothing in me but acute uneasiness and the feeling that they've all gone mad. What I cannot go along with is the men who make the same assumption about male bonding as Freud made about the penis: 'We have it, and they don't. How beautiful it is. How they must all wish they had it, too!'

Actually, no. And some of the greatest men in history have been the ones who echoed that 'no'; men who have experienced the male bond at its warmest, been embraced as a brother by church or army or establishment, and then found there was a point beyond which they could not conform. They have taken their minds back out of the common pool, and however hysterically the pack yapped and snarled around them they have resisted the pull of it because they could do no other. That I do see as beautiful.

Because the most disturbing thing about the primate male bond is that it works only when the leopard comes. And if it is rewarding for all the male cohort, it is especially so for the leader, who knows that the leopard's arrival is the signal for all rivals to cease challenging, for unruly subordinates to step into line, for all the strength of the pack to be subservient to his will. An alpha-male baboon only tastes this heady sensation at the approach of a real enemy, but an alpha *homo sapiens* with enough ingenuity can invent his own leopard. To repeat Lionel Tiger's definition: 'Males bond in terms of either a pre-existent object of aggression, *or a concocted one*.' (My italics.)

The whole business of politics and government as conducted by males hinges around the process of identifying or inventing the kind of leopard that will unite the greatest possible number of men in the tightest possible bond.

In wartime it is easy: they have territorialism and xenophobia on their side, and the other tribe, or the other nation, constitutes the leopard. But as nation-states grow larger and administration more complex, men have to be given the drive to exert themselves, and co-operate, and carry out orders, for long periods when there is no war. They subdivide into factions and cliques: the Whigs are the leopard, or the Republicans, or the abolitionists; and if the bond shows signs of weakening they bring on a sabre-toothed tiger: Popish Plot, International Jewry, Yellow Peril, Bolshevik Conspiracy, Capitalist Lackeys. It is a curious way to run a society, but this is the way men are built, and this is the psychological mechanism that powers most of their political systems.

Politically, the leopard must always be the quint-essence of evil or the system breaks down. Recently on the radio I heard one of the mildest and most decent-minded men in British politics relate how he went into a new Parliament praying that the Tories would do something really wicked. His prayers were answered: they sold arms to South Africa, so his juices began to flow again, and his job became once more absorbing. You'll admit it's a pretty odd profession, where you have to petition the Lord that your country may behave with turpitude, lest your own batteries run dry.

Many people have puzzled over the fact that even when enfranchised, few women enter political life, and of those who do, few attain high office. Lionel Tiger calculates that they never rise above 5 per cent of active politicians, and attributes this to their lack of 'bonding' and their lack of leadership. But you cannot create a bond until you can invent and believe in a leopard, and I believe that to the average woman most of these leopards are like the emperor's new clothes.

Even in wartime, where the argument is territorial and powerful—and she believes the government when it tells her the enemy is diabolical—even then, a woman actually encountering a soldier face to face is apt to see a man; where a properly bonded person would see an enemy. She may bind his wounds, instead of shooting him as she should. She may even love him, and afterwards get her head shaved for treachery.

In peacetime she is even more pusillanimous. She is liable to look at a spokesman on television and think: 'Poor old man, he's doing his best,' forgetting that she voted for the other side and therefore he must necessarily be hell-bent on bringing the nation to bankruptcy and humiliation for the most sordid of motives. Nobody with a mind like hers is fit to run a country. Or even if it should be true that she is fitter to run a country on sane lines than any of the leopard-hallucinators, you can safely bet that the way the system is geared she'll never get a chance to try. She would find the whole structure designed for a different type of mind, just as she would find a gents' urinal designed for a different type of body. The successful female politicians have already seen the light, and the leopard, or they wouldn't be where they are today.

Homo sapiens is a creature that over vast areas of the earth, in the countries and continents and centuries that the history books and the newsreels neglect because nothing 'interesting' has happened in them, has succeeded in living the moderate and mainly co-operative life to be expected of cousins of the hedonic apes. Almost all men spend most of their lives in this way. Most men spend all their lives in this way.

But because all our governance is based on male bonding, and male bonding on at best a low-powered rumbling of aggression, accompanied by the hallucinatory visions which keep that aggression alive, we are constantly in danger of seeing our communities revert at intervals into the horrible agonic semblance of a troop of baboons. Baboons, moreover, with their finger on the button that can fire off an H-bomb.

'Something respectable' is the androcentric way of viewing this transformation. But I find it hard to raise even two cheers for a mechanism that counts among its monuments the lynch mob, the Spanish Inquisition, the My Lai massacre, the blood-soaked earth of Ypres or of Stalingrad, and the obscenities of Hiroshima and Nagasaki.

It is not that women are guiltless of these things. Women in fear for their families, or their own skins, or sometimes even their property, can display as much hatred of the 'enemy' as men, and then their behaviour, whether handing out white feathers or whipping up patriotic fervour, has often appeared more ignoble than the man's, because it is not even dignified by the warmth of the male-bonded love and they are (or have been until recently) less likely to bear the brunt of the corporate violence they are helping to sanction. For all that, I believe that in the great majority of conflicts they are accessories after the fact. They may believe in the leopards, but they don't 'concoct' them.

So where do we now stand in relation to the apes and the baboons? There is clearly a strand of baboonery in human society. It may well be true that a shift in the environment had something to do with it. Recent field observations have suggested that even among chimpanzees there is a slight but perceptible difference between the behaviour patterns of forest chimpanzees and savannah chimpanzees. The latter respond to an outside threat with slightly more co-ordinated forms of aggression.

This may well have happened to the hominid. The fact remains that this behaviour does not go back in our species to time immemorial. If we accept the aquatic theory we must conclude—since the sea was at least as safe an environment as the jungle—that it evolved after we came back to the land. It is only two or three million years old, an adaptation superimposed at a late stage on the hedonic behaviour displayed by our ancestors and still displayed by our children. It proved for a while highly adaptive, in a primitive context when speed and muscle power (which are enhanced by it) mattered more than cool reason (which is hampered by it). It is adaptive no longer.

It would, of course, be absurd to claim that females have no aggression in them. Individually they may be quite as aggressive as males; and in the case of women, where they have not been culturally conditioned to a 'submissive' role, the aggression may appear more clearly. Some people even believe that if they were not taught to be submissive, and men were not taught to be aggressive, there would be no difference at all in this respect; but this is very doubtful. Conditioning may account for a lot of it—maybe 50 or 60 per cent of it—but there are two powerful reasons for believing that conditioning is not the whole story.

One reason is that if you inject a female monkey with male hormone she will behave more aggressively; and if you inject a male monkey with female hormone he will behave less aggressively. The second reason is that anthropologists studying the cultures of different tribes have found almost no occupation which isn't somewhere or other considered to be 'women's work' and somewhere else considered to be 'men's work', whether it's pottery, or weaving, or agriculture, or cooking, or even caring for the children. The one exception is killing people. No one has found a primitive tribe where women are the warriors. War is a function of male bonding.

Lionel Tiger's account of this male phenomenon is very penetrating and clear-eyed. He makes no attempt to minimize the appalling nature of its consequences in the modern world. He does try to chalk up a great score for it on the credit side by the semantic trick of using the phrase 'aggression against the environment' to explain man's positive achievement, such as crossing oceans and scaling mountains, inventing telephones, discovering penicillin. But the motive for these things is hedonic: 'Guess what I've seen! Look what I've made!' and has nothing to do with aggression. Aggression is for fighting people with.

There is one point about the baboons that he fails to bring out, though he makes one oblique reference to the research on which it is based. It is a discovery which gives some grounds for optimism, and since discoveries of this kind are in fairly short supply, it is worth examining more closely.

It has been most clearly described by Michael Chance. He points out that when a baboon is manoeuvring his way up the rank order in the hope of becoming dominant, he will never achieve his aim by a display of unbridled aggression. The baboon who finally makes it to the top is the one with the greatest power of controlling and inhibiting his aggressive instincts.

It works like this: an animal who engages in physical combat and loses the fight is statistically more likely to lose the next one. If he loses two or three in a row he has lost all hope of becoming dominant. This is the 'nothing succeeds like success' syndrome. It has been tested and proved more than once (with rats, for instance, by Allee in 1943), and under rigorous scientific conditions which rule out any chance that it only happens because the animal was a rotten fighter in the first place.

A baboon involved in an agonic encounter with another slightly higher in the rank order reacts with feelings of aggression. If he cannot control them he will fight, and the more often he fights, the greater are his chances of being defeated and losing confidence in himself. The males who get to the top are not the blindly belligerent, but those who through all the years of adolescence, and after, keep the tightest control over their aggressive reactions and use every trick in the book of diplomacy and delay and diversion to ensure that with the minimum loss of face they never actually tangle with a member of the establishment clique.

An animal of this kind is like Fabius Cunctator—when

he strikes, he strikes hard. But he has learned in his years of apprenticeship never to fight just for the hell of it, and when he gets to the top he never fights for the hell of it either. By then he rarely needs to. The important point here is that this knocks the bottom out of the popular belief that our primate inheritance is a tendency to ferocious uncontrollable rage, and we have nothing to pit against it except the pale intellectual conviction that it is wiser not to behave in this way.

If we accept that in the period on the savannah and afterward, homo sapiens acquired a behaviour pattern of male bonding analogous to the baboon's, we must be prepared to accept that for the individual at least, when he was handed an increase in potential aggressiveness, he was also handed the power to apply the brakes. Konrad Lorenz is quite unequivocal about this: 'The inhibition is an active process.... It is quite correct to speak of releasing an inhibition process.'

I find this slightly cheering. The sad thing about it is that when the male-bonded group really start whooping it up and the monkeys are 'in a state of war' the brakes appear to fail. They are a check on individual anger but not on mob violence or war. The only way of putting a brake on those would be to find some method of loosening the grip of the bond, and how much hope is there of that? Not much; but there are some signs that if it is subjected to too much strain it will, quite suddenly, snap.

It is very rarely that the image of concocted evil begins to fade in the minds of men while the conflict is still on: but something like this is happening in America as I write. Perhaps half the nation continues to behave as if North Vietnam were a symbol of evil forces which if unconquered would constitute a dire peril to the U.S. body politic and to all 'good' men everywhere. The other half has begun to ask the treacherous question: 'What leopard? I see no leopard.'

It is probably no accident that some sections of the people asking this question, especially the young, have reacted by rejecting not only the draft but in extreme cases the whole male-bonding syndrome and all its manifestations.

It is to assert virility—they don't care about virility. It is to prove courage—so who wants to? It is to achieve status and dominance and decide rank order—they reject status and dominance and rank order. It's the backbone of the political structure—they despise the political structure. It gives coherence and separateness to national and tribal units—they say they belong only to the human race. It is to defend property and family life—they renounce property and denounce family life. They have the profound conviction that the people who say they see the leopard have all gone raving mad, and they cut themselves off from the madness as an act of self-preservation by uncompromising rejection of any behaviour pattern that might conceivably be exploited to lure them back into it. They opt for the hedonic mode of the higher anthropoids, and the only people who can rivet their concerted attention are their musicians. (Mike, the wild chimpanzee Jane Goodall writes about, had the same effect on his troop, and increased his status quite wonderfully by a brand-new display technique which involved beating rhythmically on the resonant tops of old kerosene tins.)

They reject, I personally believe, too much, and at too great a cost to themselves. But they have proved at least that the agonic male mechanism is not quite the master passion in the human race that some would have us believe. The capacity for it is there, but it is an option, not an imperative, for there is still more in us of the chimp than the baboon. Unless we can design cultural curbs on aggression more effective than any yet achieved, it may be that in that fact lies our best hope of survival.

Ten

What Women Want

Freud, towards the end of his life, bewailed the fact that even after spending years trying to pinpoint it, he had never succeeded in finding out 'what women want'.

It's rather a silly question. If anyone had assembled a string of names of well-known human beings—say, Albert Schweitzer, Attila the Hun, Casanova, Gandhi, Al Capone, Einstein, Henry Ford, Peter the Hermit, Gauguin, Elvis Presley—and asked him to encapsulate an answer to the question 'What do men want?' he would not have found that too easy, either. Any answer he came up with that held true for that list would be so abstract and general that it would also hold true for all women.

But many people have a subconscious idea that women are an altogether less complex species, more like, shall we say, rhododendrons, or beans, so that somewhere just around the corner is a simple answer on the lines of 'they need plenty of phosphates', and that once this secret has been discovered life will be simpler. Women can be given what they want and they will then keep quiet, thus enabling the time and attention of real (i.e., male) people to be devoted to the important and difficult business of conducting their relations with other real people.

This concept, like the concept of an oral aphrodisiac, is a male mirage. Such a formula will never be found. All we can do is try to disentangle some of the factors contributing to the discontent of women at the present time.

To trace the fate of the female hominid and her descendants through all the vicissitudes of prehistory and history to the present day would be material for another and a different volume.

The watershed for her was the hunting-gathering era, which established the division of labour and the nuclear family. Up to that point she had been no less free, no less self-reliant, no less inventive, no less vital a contributor to economic life than the male. No one could be a mere housewife in the days when nobody had invented the house.

It was true she was subdominant, but then most of the males were subdominant also—it's impossible for everybody to be on top—and among primates the condition is not particularly irksome. A subdominant monkey may be prevented from doing something he or she wants to do—eating the best of the food or copulating with the most desirable mate—but he is never coerced into doing things he doesn't want to do. That is a later, human, innovation.

Part of the trouble was territory. Where a band of primates regards a tract of jungle as its territory, the sense of ownership is communal and the surplus aggressiveness of males is channelled into noisy skirmishes on the periphery against the males of neighbouring bands.

But once a male has established himself as the head of a nuclear family occupying a base, he begins to regard it as a mini-territory. He feels, like Tinbergen's seagull, that however small and circumscribed the area, the female belonging to him should confine herself to it, and not wander on to the territory of other males. Once he becomes a tool maker and thus begins accumulating property, he feels the property, like the female, should remain at the base; she becomes in his mind a part of the property. Whatever position he might occupy vis-à-vis other males, within his mini-territory he is the alpha, his dominance is absolute, and there is no one to exert it against except the female and her offspring.

When he graduates from a hunting-gathering economy

to an agricultural one, both the territory and the dominance become even more vital to the man. (He is *homo
sapiens* by now.) Agriculture is a laborious business and he
doesn't want to do all the labouring in the fields by himself. (In some African communities, even in the twentieth
century, the female does it all.)

His grip on the female grows tighter still as he realizes
another of her talents may now be of economic importance
to him. The young subhuman primates, however
tolerantly treated by the males, are treated by them as a
communal asset and a communal responsibility. But
with the establishment of the nuclear family the man
recognized a special relationship with the children of his
own particular woman, and with the coming of agriculture
this relationship zoomed into focus.

It doesn't matter to a communal hunter whether the
young men running beside him are his sons or anyone
else's as long as they help to bring down the quarry. But
it matters vitally to a farmer, because if they are his sons
they will labour in his fields. He had learned to persuade
horses and dogs to work for him by bringing them up
from infancy to depend on him and obey him. He could
use the same technique on people. Female cattle and
female sheep were more valuable than males because they
reproduced themselves and so increased his wealth. The
woman could do the same.

She must never leave him now. He made rules to this
effect, much tighter ones than hunter-gatherers had made.
It was God's will, he said; and the other farmers agreed.
He and the woman must be faithful to one another till
death; although it was tacitly understood that he could
put her away from him if she turned out to be barren or
in any other way substandard.

This relationship obtained throughout most of history.
Inasmuch as legally she could not leave him, nor own
property of her own, nor obtain redress for ill treatment,

nor payment for services rendered, the relationship was that of master and slave. Many women were very happy under it and loved their husbands dearly, just as some slaves were contented and devoted to their masters; but it was slavery all the same. Some of them cheated their husbands or nagged and tyrannized over them, as some servants cheat and nag and dominate their employers; but they were servants all the same.

The introduction of agriculture had one more effect: it created so much wealth that men in general could take more of their attention away from the problem of merely keeping alive and devote it to making life pleasurable. Woman, that most versatile of chattels, had her uses here also, as concubine and prostitute. Some men called down the wrath of God on such women for their sins, while those who created the demand for their services despised them for supplying it.

The phrase 'the descent of woman' had by this stage acquired a more than genealogical significance. She had descended so far that in many, perhaps most, human communities she was regarded as congenitally inferior—physically, mentally, and morally. Sometimes this conviction went so far that it amounted to what is called 'pseudo-speciation'—she was thought of as not quite part of the human race. Men would debate in sober earnest the question of whether women could be held to have souls.

Many of the beliefs were of course self-justifying. If you believe women are mentally inferior you don't bother to educate them, and as long as you don't educate them they remain mentally inferior. If you go further and make it plain that any overt sign of not being mentally inferior is unfeminine, rebarbative, and off-putting to all self-respecting males, then she will probably take steps to conceal or disguise any such blemish in herself and quell it in her daughters.

However, there is no need to rehearse here the sorry story

of her wrongs and sufferings in times gone by. Most of them have been or are being put right. So many lies about the inferiority of women have been exploded that she now sometimes believes that except in strictly physical terms there is, or need be, no difference at all between men and women. She has equality before the law (or very nearly); and she has the vote, and earning power, and education, and the pill.

Yet, in the capitalist world at least, it is in precisely in some of the countries where these new freedoms have gone furthest that men are astounded to find women most strident, most discontented, most ready to march about with banners and complain about the dreariness of their lot, and their frustrations, and about being sexual objects, and about their need of liberation.

Some people discount the movement because it appears to be led by middle-class intellectuals. But almost every successful revolution has been spearheaded by middle-class intellectuals; and this one is getting enough reverberations, at least in Western countries, to suggest that it is tuning in to a widespread and deeply rooted malaise. It would seem that a number of things very basic and important to women have been going, somehow, wrong.

I think three of these things are: their relations with their children, their relations with men, and—least important, but still a factor—their relations with other women.

The children are really at the heart of it all—certainly at the heart of all the economic problems. However fervently we demand equality, we have to cope with the fact that women are the childbearers. Some women, like Shulamith Firestone, resent even this ('it *hurts*') and would like the job to be handed over to the test tubes of biochemists at the earliest opportunity. But I'm afraid she'll have to wait quite a while for this, and meantime the childbearing factor permeates everything.

Take the demand for equal pay and opportunity. A high proportion of the resistance to this has in the past stemmed from pure pigheaded male chauvinism. Trade unionists theoretically wedded to the principle of 'the rate for the job' accepted it as perfectly natural that a woman working on the same bench as a man for the same number of hours at the same job and with equal efficiency, even if she had the same number of dependants to support, should be paid something between half and two thirds of their wages because she was a woman, and for no other reason.

Employers even more fervently supported this 'natural' order of things, and used the most ingenious arguments in support of it. It was because 'women are supported by their husbands'—yet single women get paid no more than married ones. It was because 'men have wives and families'—yet confirmed bachelors got paid no less. It was because 'women leave to get married', or 'women stay off work more often, to nurse children, etc.'—yet more than half of the female work force in Britain today is composed of married women over forty, and they are statistically *less* liable to absenteeism, to changing their jobs, or to coming out on strike, than the average man. Nobody offers to pay them at a higher rate on this account.

Yet when all the factors of prejudice and self-interest have been discounted, the fact remains that *on average* women put less of themselves into their work than men do, simply because they are childbearers and wives as well as workers. Already at sixteen or seventeen, when a boy still at school tends to put a spurt on because his main immediate objective is a career, a girl tends to do the opposite, because her main immediate objective is a home and a family. She has been taught to regard this as a full-time job. For anything up to ten years of her life it probably will be a full-time job.

Sometimes this dual role works to her advantage, but

more often, sometimes violently, to her disadvantage. For instance, a professional man who finds his job arduous or frustrating usually feels constrained to hang on and battle with his problems because otherwise he would lose face as well as income; while a married woman encountering a similar sticky patch can more credibly withdraw without feeling a failure, telling herself (as well as everyone else) that her home life was suffering, and she felt it her duty to give more time to the family.

On the other hand, a woman with young children, constrained (by poverty or the death or absence of their father) to be a wage earner as well as a mother, is likely to find herself doing two full-time jobs simultaneously, unable to let up on either; and the mental and physical strain of this can be appalling. A man, even if his wife leaves him with young children, is seldom expected even to attempt it. On occasions it has officially been recognized, by means of a 'housekeeper's allowance' that his plight is more than flesh and blood can stand. Male flesh and blood, that is. Women, it seems, are made of sterner stuff.

Because of these and a few thousand other anomalies some women now take the line that they will never be free until they get the children off their backs; just because they produce the babies, there is no earthly reason why they should actually have to *rear* the messy things; that it is a nasty menial excruciatingly boring job, and why should they, with their fine intellects, have to get lumbered with it?

We have observed a great number of unique things about *homo sapiens* and his mate, but nothing, anywhere, as unique and apocalyptic as this *cri de coeur*.

Throughout the whole animal kingdom—and in mammals above all—if there is any answer at all to the plaintive question of 'What do females want?' it is 'They want their children.' Not that they enjoy parturition; not that they desire status symbols, or heirs, or company in their old age; but that once offspring appear, messy or not, they *want* them, as unequivocally as they want food. To take a simple example, we have seen in connection with female rats that their appetite for sex may keep them pressing their levers past the point where a male would have given up. But their appetite for acquiring babies is more voracious still, as has been proved by delivering one newborn rat in response to every signal. Given the strength and stamina, a female rat would go on pressing that lever until the whole cage was knee-deep in infants.

Among primates the drive remains very powerful, though motherhood, like sex, has partially moved into the head. Thus the right way to go about mothering, like the right way to go about copulating, has to be learned, and apes brought up in isolation are liable to make a mess of it the first time unless they've seen it done. But there is no doubting the pleasure taken in the relationship. In several species—including the untender baboon—the response to a newborn infant is universal. Not only does the mother's status shoot up; other females cluster around, making submissive gestures, hoping she will allow them to take the infant from her for a while. Even alpha males, though without the submissive gestures, approach her with the same intention. But she retains the right, and the determination, to take it back from anyone if it cries.

It is very hard to believe that the propensity to take pleasure in children has in itself weakened. Just as it is impossible for an animal to inherit the defect of sterility from either parent, so it is next to impossible for it to inherit and transmit an absence of maternal instinct, because it would very rarely live long enough to do so. Pupil-reaction tests, as already noted, show that subcortical response in women to the sight of a baby is instantaneous and powerful. In primitive societies the pleasure remains undimmed; and in most civilized ones, too. I remember in my own childhood in a Welsh valley when a mother took a new baby around to friends and relatives and neighbours to show it off, the behaviour of the women was as purely primate as the behaviour of their children in the playground. There would be an involuntary chorus of 'Aw!' at the first sight of the baby's face emerging from the shawl. Someone would always say 'Can I hold him for a bit?' The mother would graciously grant the boon, but would take him back if he cried.

Occasionally we read of one of the very few crimes which only females ever commit, when a respectable woman is charged with stealing a baby from a pram, prompted by an appetite as all-obliterating in her case as lust is in the case of the rapist. Personally I have never been tempted so far. It is true, all the same, that if a live human being under the age of about three comes unexpectedly within my range of vision, I find on checking up that my eyes are clamped on him as rigidly as the eyes of a randy male would be on a topless blonde, and that very much the same daft sort of grin is apt to be creeping over my face.

How come, then, that some liberationists groan and growl when they hear people croon about the joys of motherhood? Has a yuch reaction really crept into this relationship, like the Victorians' yuch about intercourse? Having more or less exploded the myth that some women are frigid to the delights of sex, are we now to discover that some are frigid to the pleasures of maternity? Why is it that they are ready to hire other women to bring up their children, whereas even the most *spirituelle* have seldom contemplated hiring other women to relieve them of the equally earthy chore of sleeping with their husbands?

There is a vague idea floating around that only dimwits can really enjoy bringing up babies because their conversation is so repetitive and they are, let's face it, little better than animals. Agreed.

Only what's so terrible about being an animal? Most people who read Gavin Maxwell's *Ring of Bright Water* can share his intense delight in his friendship with the otters; they were so fascinating to watch, so unaffected, so confiding; in some ways almost human; at first touchingly dependent on him, then wandering farther afield but coming back of their own free will. Nobody could doubt the real and personal nature of the relationship he built up with them, or how rewarding it could be to build a relationship like that, even though they made his life immensely more complicated.

Surely nobody felt the author must have a very limited mind to take delight in the creatures. Yet they were horribly destructive; their utterances were very stereotyped, and their repertoire of behaviour patterns far smaller than that of any three-year-old. Why then do people think of child-minding as a kiss of intellectual death?

Let's go back to the hominid, who enjoyed the job, and try to imagine what was in it for her.

Well, there was physical contact—what has been called 'maternal massage', though it wasn't only maternal, it was reciprocal; and it was continual. This kind of stimulation is pleasurable to almost every kind of sensate animal. A tame owl, if you stroke its chest and then withdraw your finger, will put out a talon and grasp the finger and draw it blissfully back where it came from, hoping for more of the same. Even something as torpid as a crocodile will swim to and fro quite actively to obtain an electrical massage of the surface of its skin. So the contact was nice for both of them.

There was, for the baby, constant physical reassurance and a sense of security; for the mother, enhanced status. She had the pleasure of watching its movements, and everything it did seemed clever. When it moved away from her there were frequent eye contacts—which psychologists now recognize to be powerfully satisfying in themselves.

Above all she obtained from the baby what William Blake said men and women require of one another: 'The lineaments of gratified desire.' When it was hungry she alone could feed it; when it was frightened she alone could reassure it; when it was cold she could warm it; when it fell down she could pick it up; when it showed off she could applaud it. She had nothing else to do all day apart from a couple of hours' gathering, and she soon learned to do that with the baby on one arm, or, later, slung in some variety of papoose-holder. She was the answer to all its frustrations, which made her feel benevolent and superior, in a world where most of her conspecifics made her feel inferior. Naturally it loved her without reserve; it is pleasant to receive and easy to reciprocate that sort of love. Later, when the child got too big and demanding or when she had a younger one to carry, she and the first one sometimes went through a period of getting cross with one another; but then it would join some equivalent of the juvenile cluster and the relationship would move on to a less close but still cordial plane. Anyway, that's the way it is with apes.

What went wrong is that we became civilized; and our babies didn't. We and they belong to areas of human experience removed from one another by millions of years. Wordsworth said they came trailing clouds of glory, but they don't; they come trailing creepers from the primeval jungle. Put yourself in their place.

Your mother is civilized, with nice clean clothes on, and wants you nice and clean, too. She can't carry you around all day any longer. She has shopping to do, and dinner to cook, and dusting, and vacuuming, and diapers to wash and probably bottles to sterilize. Besides, when she carries you around you dribble down her blouse and she likes it to be clean. So she puts you in the pram or the crib for most of the time—nowadays because she hasn't got much option, though her mother may have done the same thing to her *on principle*, because the experts told her that it would ruin a baby to pick it up every time it cried.

I don't say she should revert to the papoose-carrier; I don't say that you and she have now become merely good friends; only that already the affair between you is distinctly more platonic than it was, or than you would like it to be.

A few years later, when you're about two, the generation gap widens. You come toddling through your jungle, staggering against the odd sapling or trailing liana, plucking at a leaf and hoping for a beaming eye contact; but she is under the delusion that these obstacles are the legs of an occasional table or the flex of the TV and that you've torn up Daddy's *Times* that he hasn't even *read* ... You get an eye contact, but it's full of anguish, not applause. Sometimes, if she's intellectual and catching up on her reading or overworked and worried sick about the rent, it's not too easy to get an eye contact at all except by kicking up hell. So you *do* kick up hell. Anybody would. And then she begins to look martyred.

It is her duty to drag you through a few million years of civilization in the course of one short infancy; so that though she may be the answer to some of your frustrations, she is in fact the source of most of them. Her praise is capricious: it is limited to those of your actions which don't damage her environment or annoy other adults, and this sort of behaviour doesn't come naturally to you. You can't help feeling a little bit ambiguous about her, and she can't help feeling a bit ambiguous about you.

You enhanced her status for a brief period when you arrived, and everyone congratulated her; but whether her status is sustained by her career, or her clothes, or her mind, or her spotless kitchen, or money, or mobility, or serene deportment, or hospitality—on every one of these counts your presence makes her feel subtly more plebeian, more unprofessional, more rumpled, less able to concentrate, more disorganized, poorer, more housebound, more harassed, less able to keep her friendships in repair. If there are two or three of you, the process steeply escalates.

To compensate for all this, the love between you has to be of a high quality. Sometimes it is. But it doesn't remain for long as unalloyed as it did for the hominid. Another couple of years and you're coming home from kindergarten dropping hints about toys and treats the other kids have, and you haven't, or not yet. Where she used to be virtually your sole source of knowledge and information and standards, you now pick these things up from teachers and television and perceive there are other ways of looking at things; very few children nowadays would assume Mom is more likely to be right than the mass media. Instead of being the one person who thinks her perfect, you begin at a tender age to appear in the light of a critic on the hearth, and unless she is a very unusual woman she needs that like a hole in the head.

It's not that such a woman has become denatured and gone off children. But if you take our mythical man with his eye on the topless blonde, I believe you would find that if his footsteps were dogged by two or three topless blondes who hung around his office all day, demanding constant attention, quarrelling, 'helping' with the bookkeeping, following him to the lavatory, butting in on conferences, criticizing his methods, and every five minutes wanting help with their zippers and admiration of their knitting, and if they never, never, went away—then after five or six months you would have a man less than rapturous about topless blondes. It wouldn't mean he had necessarily gone homosexual or frigid. (More likely he would have gone clean around the bend.)

For a long period of history this situation resolved itself into a class issue. Women who could afford it handed their infants over to other women, initially for wet nursing and general maternal massage, and then for the years-long civilizing process. The children were presented to their

mothers at reasonable intervals in a clean, tidy, and sub-
dued state, to be patted on the head, appraised, and ex-
horted to be good.

Peasant women did the job themselves, and were usually
able to go on getting active enjoyment out of it, because
they never aspired to elegance in any case. Throughout
English literature it appears as a mark of good breeding in
women to treat babies, as well as sex, with a modicum of
reserve. Jane Austen's heroines don't only treat an outspoken interest in 'beaux' as a mark of vulgarity; the
ability to appear to enjoy romping and having one's hair
messed up by other people's children is also deemed not
only vulgar but hypocritical. David Copperfield was well
received by Miss Betsy Trotwood when at an age to sit
quietly and say 'Yes, Aunt'; but the right place for the
squawking leaking infant David was in the big red welcom-
ing arms of Peggotty. And as late as Mary McCarthy, one
of her heroines is fascinated to hear from a nurse about the
lower middle classes: 'Even when they have plenty of milk,
and the doctor encourages it, they don't want to nurse;
they have the idea it's Lower East Side ... the class differences are quite extraordinary.'

So the sharpest yelps about child minding come from
upper and middle class liberationists who, as you can tell
from odd phrases in their books, have grown up subcon-
ciously assuming Peggotty to be one of the unalterable facts
of life.

Shulamith Firestone: 'Households were large, filled
with many servants....' 'In every family the child was wet-
nursed by a stranger....' (*every?*)—and even today 'women are often relieved of the worse strains by the existence of a servant class'.

Kate Millett solves the problem by: 'The care of
children, even from the period when their cognitive powers
first emerge, is infinitely better left to the best-trained
practitioners of both sexes who have chosen it as a voca-

tion.' This is superb, especially the 'of both sexes' bit, put in to disguise the fact that she is calling for Peggotty. How many men find a vocation to work in nursery schools? And on the age levels where men do come in, does she really believe the average city school, any more than the average conveyor belt, is really staffed by idealistic trained practitioners who have 'chosen it as a vocation'? Most of them become teachers for the same reason that Jane Eyre became a governess, because they have to earn a living, and this seems the least distasteful of the avenues open to them.

Germaine Greer urges the runaway wife to leave her children behind if necessary because 'he is more likely to be able to pay a housekeeper or a nanny than a woman is'. And when she suggested liberating women by the pleasure principle—'It is possible to use even cooking, clothes, cosmetics, and housekeeping for *fun*'—it just never came into her mind to include the baby as one of the fun things. The best way to deal with *him*, failing a nanny, is by a communal household, so that women can take turns in 'liberating' one another from him.

Meanwhile Peggotty herself is rapidly vanishing off the face of the earth. Where's she gone? She's a little bit better off now. She wants her own nice clothes and apartment and friends and leisure. She's begun counting her own calories and keeping hand cream beside her own sink. We're all turning into ladies; and while people have sometimes jocularly debated 'Do women make mothers?' nobody's ever bothered to ask 'Do ladies make good mothers?' because the question obviously didn't apply. Well, it applies now; and henceforth increasingly.

Personally, I believe that for the first four or five years a child is happier and thrives better if there is some one adult with whom he can establish an individual, personal, and continuing relationship. I don't believe that male prejudice and the physical fact of lactation are solely or jointly responsible for the tradition that this is a woman's job, normally best done by his mother. It matters hardly at all if his mother is replaced by a nanny as long as it keeps on being the same nanny: but this fact doesn't help us to liberate women from child-rearing. At best it could only help to liberate half of them at the expense of the other half.

I don't believe it is by any means the same for him if instead of an individual relationship with an individual person he gets one fifteenth of the attention of a 'trained practitioner' in charge of a crèche. Practitioners, however trained, are only human, and there are usually one or two of the fifteen that they can't really take to. As a child-adult relationship this is substandard, just as a female-male relationship would be substandard if a woman found herself one of fifteen wives simultaneously relating to one 'trained' male. It would not be good enough for her, and it is overfacile to assume it is good enough for her child.

I think the problem here has moved out of the realm of biology and into that of economics, and I should like to look at it from that angle. The point is this: we are talking here about a process of primary production, as basic to the country's economy as agriculture or coal mining, namely the production of people. It is possible, as we know, to have too many people, just as we may have overproduction of any other commodity like potatoes or fish—and it has even more disastrous results because it is harder to plough them back in or throw them back into the sea—but every society recognizes in a thousand ways that the one thing it can't altogether dispense with is people, properly processed competent people to take their turn at operating all the intricate processes of production and administration when the present operators get past it.

From the age of five onwards, society is willing to pay through the nose to get this job done. It lays out vast sums in primary, secondary, and higher education to make sure that a child who enters the system able to sit upright, and stand, and walk, and understand words, and speak, and control its bowel movements, and dress itself, and not too often run amok, will go on from there to learn to read, and write, and count, and contribute something to society.

It very rarely recognizes that the work already put in on those five-year-old children, if evaluated in economic terms at present-day wage rates, would amount to a whacking great chunk of the gross national income. If it is left for any reason with a newborn babe on its hands—'in care' —the cost to the community of caring for it is alarmingly high and rising higher every year.

I do not suggest that the contribution made to society in this way should necessarily be evaluated and paid for. But the fact that the occupation of housewife and mother is *not* economically evaluated—and it is pretty well unique among occupations in this respect—has had far-reaching consequences.

Think for a moment about a man who goes to work in an office. People pay him for what he does and have therefore expended considerable thought and planning on how and where he does it. He may spend an hour or more travelling to work by car or train. He arrives in a section of the city where whole blocks of houses have been bulldozed and high buildings erected to create a vast acreage of specially designed working space. Even if he is only a minor cog in the business machine and not earning very much money, care has been taken to see that he is enabled to work with the minimum of distraction. Those parts of his work which can be more efficiently hived off and delegated to specialists—typing, for instance—he will be relieved of. Somebody will probably bring him a cup of coffee in midmorning and later he will go somewhere and have lunch, probably with some of his colleagues. When he arrives home, he considers—very reasonably—that it would be a bit much to expect him to take on the chores of shopping and cooking and cleaning, when his wife has been there in the house all day with nothing to do except look after the baby. Yet all he has really used in the way of working equipment has been a desk and a telephone. After all the exhausting commuting designed to put him geographically nearer to his colleagues, when he has needed to communicate with one of them two floors down he has probably done it by phone.

By contrast, the job of child-rearing—an occupation equally essential to the economy—has remained to all intents and purposes a cottage industry. Suppose *he* stayed at home in the apartment with his desk and telephone, and his wife left at 8.00 A.M. to do a hard day's child-rearing, and arrived at a centre where just one local block of houses had been bulldozed to create a few acres of floor space with sandpits, and play areas, and a paddling pool, and sound-proofed sleeping cubicles, and a launderette for nappies, and a TV room for children's programmes, and a children's feeding room with high chairs, and a cafeteria for the mothers to take lunch on a shift system, and in lieu of typists a few specialists to mix the formulas and sterilize the bottles and clean up at the end of the day, the way offices get cleaned. She might then begin to feel her job was as important as his—which God knows it is—and needs at least as much special equipment, and benefits at least as much from regular contact with others employed on the same task. If she had a mind free to concentrate on it, she might even rediscover that this job is far more re-warding and creative than most, and that young children are even more fascinating to watch than otters, and that a juvenile cluster is less clinging than an isolated tot, and that when her attention isn't on a thousand other things he doesn't have to kick up hell to get a piece of it.

Perhaps it was special pleading to pick on an office worker. It he were an executive his wife would have a house in the suburbs and a rumpus room and daily help, and if he were a truck driver or a factory worker his own conditions of work might be pretty intolerable. But sooner or later somebody has to *think* about his conditions. His union will come out on strike, or in the last resort he will vote with his feet and quit his post, or take a vow that at least his sons will never go in for a job like that, and in the end his employer will find 'you just can't get people to do it'. You have to introduce pithead paths, or more machinery, or shorter hours and longer vacations: otherwise you get a discontented work force and an inferior product.

Housewives and mothers seldom find it practicable to come out on strike. They have no union, anyway. But the rumblings of women's liberation are only one pointer to the fact that you already have a discontented work force. And if conditions continue to lag so far behind the industrial norm and the discomfort increases, you will find—some people believe that in the cities it is beginning to happen already—that you will end up with an inferior product. That's going to be a very bad day for everybody.

Eleven

Present and Future

The maternal relationship, then, seems to be offering less immediate biological reward to many women, largely because the environmental context is inimical to it.

But the opposite, surely, is true of her relations with the male. The actual performance of the sex act should be, and usually is, more pleasurable to her than it has been for her predecessors over a good many generations. She has been relieved of a good deal of the load of artificial and unnecessary shame and guilt associated with it, and the amount of attention concentrated recently on her own sensations and reactions and responses has been unprecedented. One might expect her to be overwhelmed with feelings of joyous gratitude for this. The whole relationship between men and women should by now be irradiated with a cordial new atmosphere of warmth and comradeship and mutual esteem.

In individual cases I have no doubt that this has happened. But only an optimist would maintain that the net result of recent developments has been to make men and women on the whole *like* each other any better than they did under earlier, less permissive régimes. There are plenty of signs that in many ways they have actually less liking and respect and admiration for one another than their great-grandparents had in the old days when in the words of the old cliché men were men and women were glad of it; and chastity had not been outmoded; and sex was so hedged around with taboos that, as Thurber wrote, 'It got so that in speaking of birth and other natural phenomena, women seemed often to be discussing something else, such as the Sistine Madonna or the aurora borealis.'

We don't want to go back there. A lot of cant has been swept away, and the areas of human experience that could not be spoken of have drastically shrunk, and this cannot be anything but a solid gain. The only advantage of the old system was that in essentials it had been in operation for a pretty long time; people were used to it and knew where they were, and what roles they had to play, and to nine people out of ten this is always a great comfort.

The roles they played were based upon a script constructed around a few basic axioms. One was that men were created dominant and would always remain so because of their superior strength and superior wisdom, and because it was the will of God. (Milton: 'He for God only, she for God in him.') But in a secular and mechanical age Milton's God is out of date, muscle power seems to have less and less relevance, and even the male's superior wisdom is not the self-evident proposition that it once was.

Another axiom was the division of labour. Woman was unfit to face the harsh realities of economic life, so her place was in the kitchen and in the nursery. As long as there was no way out of this, most women adapted themselves to it very well, and took pride in it, and the nuclear family (based from the beginning far more on division of labour than on sex) continued to cohere. Nowadays most women for some part of their lives face the harsh realities of economic life, and find them far from intolerable. They have also discovered that male dominance was not so much based on the fact that he had more muscles and more wisdom, but on the fact that as long as she stayed in the kitchen he had *all* of the money.

An even more venerable axiom going right back to the Garden of Eden was: 'In sorrow shalt thou bring forth.' It was one of the eternal rules that any act of sexual intercourse was likely to be (inside marriage) 'blessed', or (outside it) 'punished', by pregnancy. Now, new contraceptive methods, though still comparatively in their infancy, have set a light to this one. It is burning its way along a long fuse, but the evolutionary bomb at the end of it has not yet gone off.

With so many bastions of his dominant status skidding out from under him, man hung on tight to the only symbol nobody could take away from him. He still, by God, had his penis. However cool and efficient and economically independent a female might be, if he ever had any tremor of doubt that he was worth three of her, he had only to remind himself that underneath that elegant exterior was a nude female with all the usual sexual appendages. If he was driven to ask himself what the position of women ought to be, he could always—if only in his mind—come up with Stokely Carmichael's answer 'Prone'. (I have never been quite clear whether Mr. Carmichael had a sexual prejudice against the 'missionary position', or whether he just didn't know the difference between prone and supine, but it was obvious to everybody what he meant.)

This reaction of course is not typical of all men, or even of most men. Most well-adjusted men, especially intelligent ones, have on the whole welcomed the emancipation of women, if only because they have to spend at least part of their time in the company of women in non-sexual contexts—even the marital context is non-sexual for most of the twenty-four hours—and it is less boring to talk to women since they have acquired a few more topics of conversation.

However, I think the reaction is one factor contributing to the astonishing boom in sex and pornography. The urge isn't new; it was always there, but the recent wave of obsession with it in Western countries seems to be new, and the women's liberation complaint that females are being regarded more and more as 'sexual objects' has a lot of truth in it.

Only a very small minority of women as yet are 'complaining'. Most of them are rejoicing. Sex is nice; being looked at and admired and chatted up is very nice; and the keen competition to be the sexiest among the local sex objects is worth millions to the manufacturers of cosmetics, perfumes, eyelashes, miniskirts, hot pants, and the pill.

Reactions to all these phenomena are sharply divided. Some people see the new attitudes towards sex as a tremendous liberation of benevolent life-enhancing forces once cruelly held in chains by sour-faced puritans. It is regarded by others as a Gadarene rush away from all standards of decency and morality down a muddy slope into filth and debauchery. One side sees it as an emergence into sanity and sunshine; the other as the crumbling away of the very foundations of order and civilization.

These reactions are both slightly hysterical, and accompanied by acute manifestations of mutual aggression, fear, hatred, and moral indignation, with each side totally convinced it has a monopoly of the only really moral morality.

They hurl atrocity stories at one another. One side weeps for 'the youth pined away with desire and the pale virgin shrouded in snow', couples trapped in impossible marriages, unmarried mothers pilloried by prejudice, children tortured with guilt and fear because they'd been told masturbation was deadly sin and led to epilepsy and dementia; homosexuals hounded and persecuted simply because they loved one another; Marie Stopes pelted with filth and threatened with arson.

The other side points to soaring figures of venereal diseases, abortion, and drug deaths; to shattered children of homes broken by adultery, desertion, and divorce; to schoolgirls promiscuous at eleven and pregnant at twelve; to cynical commercial exploitations of pornography and exhibitionism and perversion driving family entertainment out of cinemas and theatres.

It is very unlikely that the net effect on the total of

human happiness will be as great as either side believes. Some things become easier with greater 'permissiveness', others become harder. People are less likely now to be embarrassed when a man says he loves another man, which would once have been shocking: they are more likely now to be embarrassed if he says he loves his mother, which would once have been commendable. It is easier for a young girl to kiss a young man in public; but a recent inquiry revealed that in many areas she would be chary of walking about with her arms around the waist of another girl—though ladies in the novels of Jane Austen and Dickens and Tolstoy do it constantly with complete lack of inhibition—because now she has heard of lesbianism and it has taught her a new taboo.

Guilt and anxiety are not being dispersed, only attached to different situations. There is less shame attached to losing one's virginity too soon, and more attached to keeping it too long. It is less taboo to say 'shit', and more taboo to say 'nigger'. There is less fear that you can be unbalanced by masturbation, but a new conviction that you can be unbalanced by abstention. Less obloquy attaches to sleeping with a girl without giving her a wedding ring; but to do it without giving her an orgasm is a newly patented way of lousing up your self-esteem and peace of mind.

Tolerance is not really being enlarged: it is moving its targets. The woman who cuts loose from an unpleasant husband because she cannot bear to live with _him_ is praised where once she was condemned. But the woman who hangs on to a reluctant husband because she cannot bear to live without him is condemned where once she was praised. Anyone who succumbs to alcoholism meets with less censure and more compassion than formerly ('it's an illness, really ... perfectly understandable, the pressures are too great...') but anyone who succumbs to obesity gets short shrift ('no excuse for it these days ... only needs a bit of will power ... _other_ people manage not to let themselves go. ...') The total number of moral attitudes struck, the difficulty of trying to conform to them, and the weight of social disapproval visited on those who fail, vary hardly at all.

As for the obsession with sex itself, it is partly a by-product of affluence. Less and less time and attention needs to be given to the gratification of other physical needs, so this one is thrown into prominence. Even in primitive societies sexual activity is heightened at periods when the community is more than usually well stocked up, so that there is feasting and no need to go foraging for several days.

For people with boring jobs whose work only demands a small fraction of their mental capacity—and there are more of these every year—sex provides them with something interesting to think about; for people starved for love or a sense of identity it ensures that at least one person will pay them close attention for a while; for those who win the rat race it is a trophy and for those who lose it, a consolation prize.

The trouble is that sex as a pastime, when divorced from love, has one serious drawback. Like many forms of physical gratification, it is subject to a law of diminishing returns. To a hungry man any food is delicious; to a not very hungry man only delicious food is delicious; to a sated man no food is delicious. It is very frustrating for a man with the means and the opportunity to satisfy an appetite when he finds the appetite itself is failing him. The Romans got so frustrated that they built vomitoria where they could go and empty their stomachs and come back and eat again.

In some of its more extreme aspects the sexual revolution seems to have passed the point of campaigning for the liberation of a natural appetite, and reached the vomitorial stage of trying to reactivate an exhausted one.

Up to a point, as any biologist knows, it is possible to

achieve this. When a given stimulus, on account of repeated applications, ceases to elicit a given response, it is possible to reawaken the response by increasing the stimulus. The foster parents of the cuckoo's chick work themselves to skin and bone to rear their enormous changeling and let their own go hungry, because a large gaping beak is a stronger stimulus than a small one. Many birds will show a preference for trying to hatch a larger-than-life egg; a male butterfly will get besotted over an artificial female with larger-than-life spots on her wings.

In terms of human sex this technique can be applied in various ways—cosmetic aids can supply redder lips, longer lashes, brighter hair, whiter teeth, larger breasts, or smaller waists as fashion may demand. There are, however, certain natural limits. Where the demand for increased stimulation centres on increased exposure it runs into a cul de sac, because you can't get nuder than nude. Once full frontal nakedness as a public spectacle has become another déjà vu, there is no further for it to go except into the nightmare of one cartoonist who drew a stripped striptease girl responding to the demand for more by gracefully, with an enticing smile, drawing out her entrails and displaying them to her avid audience.

Recently there have been some signs that the sex boom is running out of steam, and certainly in some areas it is encountering a vigorous backlash. Much of it has been due to a well-recognized syndrome known as 'cultural shock'. At least it is recognized by anthropologists, who know that primitive tribes have sometimes literally died of it. But at home many progressives, who are hotly indignant when ham-handed imperialists trample over the taboos of subject races, fail to see that their sexual iconoclasm is inflicting the same trauma on some of their fellow countrymen. This doesn't necessarily mean the process should be halted. It does mean it should be carried out non-aggressively, and the words 'I believe you are shocked!' should be spoken not with derision but with concern, whether the shock was inflicted by defective electrical wiring or by a change in sexual mores.

How will all this finally affect the status of women? They will have some hard thinking to do and some careful adjustments to make if they are not to end up losers on the deal. Because what is happening is a slackening of the rules. Some of them were bad rules; and I believe they will inevitably be replaced by new rules, because that is the nature of human society. Meanwhile, whenever you get a situation where the rules are temporarily suspended— as in the Wild West before the lawmen came—the effect is that the tough come to the top and the weakest go to the wall. And women, in the aggregate, are not the tougher sex.

Thus one effect is that there are rather fewer sexual problems for young men, as chastity gets outmoded; but a higher proportion of young women are faced with the still formidable crises of unsupported motherhood or abortion. Insofar as it is true that more men are content with casual sex and more women desire a permanent relationship, the males are now capturing the moral initiative; so that if a girl does want love and marriage, she can now sometimes be conned into actually feeling ashamed of wanting them, and denying with profuse apologies that she had any such unreasonable thought in her mind.

Males are capturing the hypochondriac initiative, too. In the old days it was the bride who had to be treated tenderly, with infinite tact and patience, if the relationship was to be a success. Now it is the groom whose delicate ego must be cosseted, because he has a more fragile piece of machinery there than was dreamed of in the old philosophy. To judge by the letters sent in to some male magazines, he spends half his life worrying because his ejaculations come too quickly or too infrequently or in highly specialized circumstances, just as mothers used to worry about similar aberrations in their babies' bowel movements, until Dr. Spock breezed along and posed the cosmic question: 'So what?'

Despite all this there are some women's liberation types who are in the forefront of the sexual revolution and calling for more, on the grounds that marriage can be slavery, and sex is getting more democratic, or on the more general grounds that things have been so horrible up to now that they want to change everything. However, these are for the most part pretty tough babies who know they will survive even the most drastic upheavals. And even they don't find it too easy, because a sex-ridden society is always ready to resurrect the old slogan of 'woman's place is on her back', and a man whose gaze is too avidly riveted on a woman's cleavage only gets irritated if he's asked to listen, really listen, to any words coming out of her mouth.

It seems to me that for women particularly, life in a sex-obsessed society has one very depressing drawback: it alters the graph of her life in a peculiarly debilitating manner. To explain what I mean, flash back to the primate for a moment.

Among the anthropoids, sex appeal still comes to everyone like manna from heaven, as a regular event. Every non-pregnant female takes her turn at being the most desirable of all her troop. And that cycle keeps turning for as long as she lives. A female baboon has no menopause as we have; and not only does menstruation not stop, oestrus doesn't stop either, nor is there any law of diminishing returns. She may grow old and skinny, but the time of the month still comes round when she and nobody else is the sexiest female of the whole troop, and sends them crashing through the ceiling just as effectively on her last cycle as on the first one she ever experienced.

It was not so with the hominid. The biological emergency had been survived; sexual relations had settled down again, made slightly jumpy by the emotional hangover of conflict, and slightly puzzling by the infusion of emotions of love into a once casual physical transaction. But oestrus was gone forever.

Female desire and female attractiveness, instead of looping up and down once in every four weeks, described simply one long slow parabola. Around puberty the female hominid came into bloom; her desirability climbed, and held throughout her prime, and then declined. After that her interest in the males diminished, and theirs in her.

In prehistoric days there is no reason to assume that the inexorable fading of her charms caused her any distress. When living conditions are fairly rigorous, sex as a subject to worry about comes a long way down the list, after food and water and enemies and wild animals and evil spirits and sick children and trying to keep dry in the tropics and warm in Tierra del Fuego.

Also, after *homo* became *sapiens*, age itself brought honour. One of the most vital factors in human evolutionary success was the power to accumulate knowledge, to profit not only from personal experience but from the experience of others, even of others long dead. Before the invention of writing this was made possible only by the long life and memory of older members of the tribe.

When something 'unprecedented' happened—a flood, an epidemic, a plague of locusts—old men and women who had seen it all before could look back fifty or sixty years and 'prophesy': 'The water will rise no farther than that rock'; or 'Many will sicken but few will die'; or 'If you do this, it will be of no avail'. Today things change so fast that the experience of the last generation is increasingly irrelevant, and the bottom has dropped out of the market for venerable sages of either sex.

Since the subject of the Older Woman has now cropped up, this is as good a place as any for a short détour on the menopause. It is yet another of those biological phenomena unique to our species which seem perfectly easy to explain until you really start thinking about them.

The 'obvious' explanation of the menopause is that after a certain age it becomes more dangerous and harmful to a woman to conceive, bear, and raise children; and therefore the benevolent forces of evolution protect her against this danger by making conception impossible. But this is not the way things work. As we have previously noted, the forces of evolution have no interest in benevolence towards the individual.

Gestation and nurture may become equally arduous and debilitating for an ageing chimp or an ageing gorilla, or for that matter an ageing cow, but in no other species is the female biologically compelled to retire from these duties in order to prolong a serene and untrammelled old age.

In the same way, it is very dangerous and harmful to a leopard for it to lose its teeth; but natural selection will never militate against toothlessness in old leopards, because once such an animal is past its breeding prime, its individual longevity is of no further advantage to its kind. It is only consuming resources that would otherwise be available to its progeny, and the sooner it shuffles off this mortal coil the better it will be for the progeny and the species.

The only way of accounting for the evolutionary emergence of the menopause in women is by the assumption that the tribe as a whole, and not merely the individual, derived some benefit from the presence of those females who although sterile lived to a ripe and healthy old age. In some way or other, and in a way that applied to no other species that we know of, grannies were good for them.

As far as I can see the explanation of this can only be found in their function as repositories of 'wisdom', as described above; and especially as it related to their particular craft, the care of the young. There was no Doctor

Spock in those days. The best way of treating, for example, a child with a broken leg could only be discovered by trial and error, and in the case of such infrequent hazards only someone with a long memory would be likely to recall a time when it had been handled rightly and another time when it had been handled wrongly, and compare the results, and draw the inference.

True, there were men around, not driven to an early grave by child-bearing; but their attention tended to be concentrated on other concerns. Old women were repositories of child lore as old men were repositories of hunting lore; so it was adaptive for the species that annual parturition should not continue to the point where it would have drastically shortened their lives. Any group in which a menopausal mutation occurred and became established would be fitter to survive than groups in which this had not occurred.

Thus women alone among primates evolved the menopause because they alone among primates had acquired a method of furthering species survival that had nothing to do with their wombs. They could remember; they could think; and they could communicate their memories and their thoughts.

It can be a little inconvenient (especially to daughters-in-law) that the instinctive prompting to interfere and 'know best' in such matters still tends to well up in the middle years in a society where it is often irrelevant; but these days most of us are rigorously trained to inhibit the impulse. On the other hand, when some of the more rabid male chauvinists try to define us exclusively in terms of our hormones and our internal plumbing, it is good to reflect that millions of years ago the blind impartial forces of Darwinian selection registered the fact that the value of a woman cannot be computed by assessing her only from the neck down.

To return to the new graph of sexual attractiveness:

Up to a few generations ago the decline of sexual attractiveness was still not hard to take. The change of role from blushing bride to full-time mother happened in a few short busy years and most women after the first nine or ten pregnancies would be uttering fervent prayers for the whole business to be over and done with. Even Queen Victoria, a loving wife if ever there was one, grew to feel strongly that one could have altogether too much of a good thing. There would be all the children to be absorbed in and worried about, and then the arrival of the grandchildren, and then good night.

The shape of our lives is vastly different now. The children are fewer. They need *economic* support for a longer period than ever; but the actual physical chore of supervising the average two-point-something offspring after they have reached school age is simply not enough to absorb the energies of their mother for the rest of her (greatly extended) active life. As for grandmotherhood, which used to mean a resurgence of importance in a new, pleasant, and well nigh indispensable role, it is not what it used to be, certainly in the West. In a society where sex is king and youth at a premium, a forty-two-year-old granny has mixed feelings about laying claim to the title, and with more mobile populations and the fragmentation of the extended family it is a relationship increasingly conducted at long distance via phone calls, and birthday cards, rather than in the chimney corner with fairy tales and lullabies.

The net result is that a girl who plumps joyously at sixteen for being 'strictly a female female', with her eyelashes all in curl, and her sights trained on the 'career' of marriage, embarks on adult life looking sexy, having fun, and with everything going for her. Anyone who approaches her then and says, 'It's all very well being beautiful, but keep pegging away at your maths because you may need it yet,' or 'What about equal pay?' is going to get a very short answer. She knows that youth's a stuff will not endure, her status is as high as a baboon's in full oestrus, she's hell-bent on falling in love, and being fallen in love with, and living happily ever after. Nobody can blame her. It's the way she's been conditioned to think.

Around thirty-five or thirty-six she looks over her shopping list one week and sees, with a comic ruefulness, that it includes a couple of items like anti-wrinkle cream and a new slightly more supportive foundation garment because a body stocking no longer quite fills the bill. Slowly, consciously or subconsciously, it gets borne in on her that from here on, for the strictly female female in a sex-obsessed society, the role gets tougher all the time.

This is where, in the more prosperous sections of society, all that famous neurosis begins to set in. If her husband is in the rat race she doesn't dare let up on looking sexy, because her husband's image suffers considerable damage if his wife isn't at least trying her best to be sexually attractive. It used to be okay if she was faithful and patient and competent, but now he has this thing about his virility, and the most sure-fire way of proving it is to have a woman in tow who makes the other chaps feel: 'Boy, he's doing all right for himself there!' Moreover, marriage isn't as binding a bond as it used to be. If he feels she's seriously letting him down in this department, he's liable to look elsewhere for this status symbol, and possibly think about switching over in early middle age to a Mark II wife maybe ten or twelve years younger. Because the graph for a man doesn't follow the same curve. His status (sexual as well as social) depends to a much greater extent on factors that at thirty-five or forty are still on the upgrade—power, and know-how, and money.

America is the place where these attitudes first appeared; they are not nearly as prevalent outside it. It is also the place where (no accident) women's liberation first began to make a real noise. And for most of the Western world, for good or for ill, it seems to be the place the wind blows from where social changes of this kind are concerned. If they are, as they appear to be, consequences of increasing affluence and increasingly detaching the concept of sex from the concept of love, they are likely to spread.

'Matriarchy' is a word often applied to American life, but one of the best comments on this came from J. B. Priestley:

'If [American] women become aggressive, demanding, dictatorial, it is because they find themselves struggling to find satisfaction in a world that is not theirs. If they use sex as a weapon, it is because they so badly need a weapon. They are like the inhabitants of an occupied country. They are compelled to accept values and standards that are alien to their deepest nature.... A society in which a man takes his wife for a night out and they pay extra, out of their common stock of dollars, to see another woman undressing herself is a society in which the male has completely imposed his values.' Woman 'is compelled to appear not as her true self, but as the reflection of a man's immature, half-childish, half-adolescent fancies and dreams. Victorious woman forms a lasting relationship with a mature man. Defeated woman strips and teases.' If these tendencies continue to spread we shall all be facing defeat.

No one can go on about a problem at the length I have been going on without raising the expectation that the last chapter will demand in ringing tones: 'What then must we do to be saved?' and come up with a slick answer. Anyone who fails to do so may be accused of chickening out. I haven't got a slick answer, and I don't particularly mind being accused of chickening out. But since there are a few things I feel quite strongly we ought *not* to do, it might be a good idea to take a tentative stab at considering where we might go from here.

What we surely mustn't do is try to found a women's movement on a kind of pseudo-male bonding, alleging the whole male sex to be a ferocious leopard, and whipping up hatred against it. We mustn't do this for four good reasons.

1. In the words of Bertrand Russell: 'To love is wise: to hate is foolish.' Any damage it might do to the hated is nothing compared to the corrosive effect it has on the hater.

2. It is arrant nonsense to pretend that men are hateful. Not more than 2 or 3 per cent of them are activated by malice against women. It's just that while things are in a state of flux they are just as confused about their role as we are about ours; most of them, if they see any advantage to be gained from the confusion, will attempt to cash in on it, and most women given the chance will do the same. It takes two to tango, and it takes two to make a woman into a sex object: at the time of going to press most women are highly flattered to be so regarded, and would be insulted if their efforts to look sexy weren't rewarded with precisely this 'tribute'. If some women feel trapped by marriage, you can bet your bottom dollar that at least as many men feel trapped by it, and any woman feeling disenchanted by the status quo should pay heed to Thurber's heartfelt answer: 'We're all disenchanted.'

3. As a bonding mechanism it just won't work. Most women don't hallucinate that easily. You may raise the alarm and beat the drum, but when you point your finger at the enemy, most of them will say: 'No, no, those aren't leopards. That's the postman, and that one is my son, and the one with the nice blue eyes is the one who was so kind to us last winter when there was all that snow.' And they will be right.

4. Where a bonding mechanism doesn't work, more than half the steam that's been worked up gets diverted from the 'enemy' and redirected against the 'traitors'. This we just can't afford. Most women have far too little self-confidence anyway, and when they start criticizing one an-

other everything gets ten times worse. The non-working wife gets on the defensive because she feels the working ones think she's turning into a vegetable; the working ones are on the defensive because they feel the full-time mothers think their kitchens are in a mess and their children neglected. Childless women write defensive letters to the papers, feeling they are being called selfish because they'd rather have their freedom and a new car or go on with their careers; mothers of five are on the defensive about the population problem. It is time we stopped all this nonsense.

The first of all the things women need to be liberated from is their chronic tendency to feelings (admitted, concealed, or aggressively overcompensated for) of guilt and inadequacy. A woman who feels bad because her house in in a mess is tempted to restore her self-esteem by sneering at her house-proud neighbour: but what on earth is wrong with being house-proud if that's what turns you on? Keeping a house beautiful is no more barren or 'stultifying' a job than a professional gardener's keeping a garden beautiful.

Any attempt at 'bonding' women into a cohort all facing one way is not only doomed to failure, but will result in undermining their self-esteem still further. What is called for is a return to the invaluable primate female habit of frequent and assiduous mutual grooming, translated (as most grooming behaviour is these days translated) into conversational terms.

Much of it, of course, still goes on, largely over the telephone. Most men contemplate it with bafflement, because the actual informational content of these chats is often minimal. But information is not the point of the exercise: it serves the need, still basic in most women, to lessen the feelings of isolation by friendly and reassuring interaction with someone of the same sex. To say 'But you could have told her all that in two minutes and anyway you'll be seeing her on Friday' is as stupid as evaluating the benefits of anthropoid grooming by counting the number of ticks extracted.

The only thing wrong with it is that it is too often limited to interactions between women with a similar life-style—mothers of young babies, or widows, or divorcées, or career women; mothers with teen-age daughter problems, or teen-age daughters with parent problems. Within such groups they often find it heartening to get together to boost one another's morale when things are falling apart. Anyone who really wants to improve the status of woman as a whole should make it a point of honour to send the same friendly signals *across* these frontiers also, when occasion arises.

Out with the hate bit, then. I admit to feeling uneasy on this account about Kate Millett's *Sexual Politics*, as well as a few other liberationist writings along the same lines. It's a highly intelligent book meticulously analysing the pornographic fantasies incorporated in the works of some high-rating and best-selling male authors. But what is Kate Millet's book *for*? What it is saying to women seems to be something like: 'This is what men really think of us. It's pretty loathsome and insulting stuff. We do right to hate them.'

I doubt it. I doubt whether this kind of writing has anything to do with politics, or with anything at all in the real world. Without having met the gentleman I would hazard a bet that not even Mr. Norman Mailer actually moves around the United States committing brutal sexual attacks on casually encountered females. Surely this is dream stuff, male soap-opera, and the women in it are dolls, not people. And in their waking moments the men who write it must be aware of this truth and act on it; otherwise they would be certifiable.

Let us grant that men, or some men, have some of this stuff fuming around in the bottom of their minds. It's been left there from a very long time ago; it's a little surprising it hasn't evaporated yet. But it has no more 'political' significance than Jack and the Beanstalk. It shouldn't be too hard to verify this, for some women likewise have masochistic fantasies, and doubtless they form pair bonds with 'sadistic' dreamers and play bedroom games together, as dramatized by John Osborne in one of his plays. The sixty-four-thousand-dollar question is whether the 'submissive' partner in these capers is necessarily any more likely on that account to give way in the cold light of dawn over the colour of the new drawing-room carpet, or anything else she feels strongly about. And I suspect not. Any more than the jackbooted 'governess'-type prostitute could flog an extra thousand dollars out of her kinky clients with her whip. Dream worlds have no effect on where the real power lies.

If we don't go for hate, what should we go for? Two or three objectives seem fairly clear. First, as for any other ex-subject population, greater self-respect. I remember watching Pierre Trudeau in a confrontation with a group of young females, and he addressed them as 'girls'. They informed him that they had recently attained their majority, and so were no longer girls. This shook him somewhat and he floundered for a minute looking for another polite euphemism for what they were. 'Er—ladies?' he hazarded. 'We are women,' they said, as if they were proud of it. It was like the first time somebody said right out loud: 'Black is beautiful.'

Second, economic independence; because until every woman feels confident that she can at need support herself we will never quite eradicate the male suspicion that when we say, 'I want love. I want a permanent relationship,' we really mean, 'I want a meal ticket. I want you to work and support me for the rest of my life.' It needn't mean the end, for everybody, of the division-of-labour family. If a man wants a wife who will stay home and raise his children and finds a woman who wants to do just that, then that's fine; as long as he has paused to assure himself that it *is* what he wants and that like anything else it costs money; and as long as she has paused to ask herself the important question, 'First I will raise our children—and *then what*?' Because the 'then what' may last for forty years and she doesn't want it all to be anti-climax.

Third, the *certainty* of having no more children than she wants, and none at all if she doesn't want any. This is essential not only for women but for everybody, because every human being should have the inalienable right not to be born to a mother who doesn't want him. Once this is fully achieved it will be in the power of every woman to decide whether or not she wishes to have a child.

'Take what you want,' said God in the old proverb, 'and pay for it.' If she wants this, one way or another she's got to pay for it, and it doesn't come cheap. She may do it by devoting a few years of her life to rearing it. She may do it by settling for at least a period of economic dependence (probably on a husband) while she's doing it. If she's very fiercely independent she may do it by settling for a period of comparative economic penury while she's doing it.

She may wish to have the child and get someone else to do most of the rearing, and this is fine if she's lucky enough to have sufficient capital or earning power or a rich enough husband. She has the right to shout and complain and move heaven and earth to try to get some public recognition that the job she's doing is important to society, and money should be expended on enabling her to do it better and more efficiently; and to combine with other women to set up play groups or anything else that will make things easier until such time as heaven and earth begin to listen to her. She has the right and duty to select a husband who also wants children, if she wants them herself; and to urge him to help her as far as he is able and willing.

What she will no longer have any right to do, once 'accident' is altogether ruled out and every child is the result of conscious choice, is to give birth to it and then shortly afterwards start raising the cry of 'Will no one for Pete's sake come and take this kid off my back?' If we campaign for more efficient and foolproof contraception and free abortion on demand (as I believe we must), then we must face the moral consequence of this, which is that motherhood will be an option, not an imperative; that anyone who thinks the price too high needn't take up the option; and from that point on, where children are concerned, more inexorably than ever before, the buck stops here. On the distaff side. If we try to dodge that, we lose all credibility.

What about marriage? The more way-out liberationists seem to be bell-bent on destroying the institution. I can't quite see why there has to be a 'policy' about this. When we're just getting loose of one lot of people laying down the law that we *must* get married, it's a bit rough to run head on into another lot telling us we mustn't. It is surely, as Oscar Wilde ruled about the tallness of aunts, a matter that a girl may be allowed to decide for herself.

Anyway, marriage is going to be with us for a long time yet. As Shulamith Firestone mourned: 'Everybody debunks marriage, but everybody ends up married.' And one of the most durable statements ever made about it was Dr. Johnson's: 'Marriage is not commonly unhappy otherwise than as life is unhappy.' It can sometimes be tough for two people of opposite sexes trying to live permanently at close quarters without driving each other up the wall. But it can be equally tough trying to do it with someone of the same sex, or with a child, or a parent, or a sibling, or a colleague; or with a succession of different partners; or with a commune (for the rate of failed communes is at least as high as the rate of failed marriages). And it can be toughest of all trying to live in an empty house or apartment quite alone.

Nor is there much fear that men, once sex is more freely available, will seriously seek to escape the 'trap' of matrimony. Even on the physical level, there's nothing quite like having it on tap at home, without having to go out in all winds and weathers to chase after it. Besides, though they seldom admit it, their psychological need of a stable relationship is as great as ours, or greater. After studies carried out at the Mental Research Institute in Berkeley, California, a research group reported: 'In accordance with the popular idea of marriage as a triumph for women and a defeat for men ... we could expect to find those men who escaped marriage to be much better adjusted than those women who failed to marry.... The findings suggest the opposite. More single men are maladjusted than single women [as shown] particularly in indices of unhappiness, of severe neurotic tendencies and of anti-social tendencies.'

So marriage (or something less legalistic but the same in essence) will certainly endure until the people who say it's a miserable institution can come up with a convincing answer to the question, 'Compared to what?' I haven't been convinced by any of the answers yet.

But can marriage (or even sex) survive, once women have achieved equality and independence? The cichlid school of thought affects to have grave doubts about this. The cichlid is the fish that the 'psychological castration' boys go on about. It appears that a female cichild is incapable of mating with a male one unless he is aggressive, belligerent, and masterful; and a male cichlid is rendered impotent by a female who fails to put on a display of timorousness and subservience. Therefore, it is subtly implied, if women ever attain equality, then we will find to our horror that men are no longer men, and we will all heartily wish that we hadn't been so hasty.

What we are less often reminded of is that human beings are not fish, but mammals; that psychological castration is quite a common feature in many mammal societies also, but that in these cases the mechanism is totally different. In the vast majority of mammal species the only creature who can psychologically castrate a male mammal is *another* male mammal; and he does it quite simply by beating him in fair combat. This pattern is exemplified over and over again in studies of primate behaviour, but the most classic and frequently quoted illustration of the process comes from cattle. There was this bull who was growing older and no longer able to service all the cows in the herd, so they brought in a couple of younger bulls to help him with the chore. He challenged them; he fought them; he defeated them. And not only were the defeated bulls psychologically castrated, but the victorious one had attained such an access of virility that he returned to his harem, serviced all his remaining wives, and snorted around for the rest of the season like Alexander looking for new worlds to conquer. Any man who insists on playing the cichlid game and complaining he's castrated because the little woman isn't being submissive enough shouldn't be surprised if she asks him what's going on at the office lately.

For the real answer to this we needn't go to the animal kingdom at all. In Soviet Russia, women have had economic equality for a long time now. Seventy-five per cent of their doctors and teachers are women, and 58 per cent of their technicians and a third of their engineers, and 63 per cent of their economists, and nearly half their scientists and their lawyers, and all the women in all the jobs get equal pay. And while I have heard a lot of critisisms levelled in the West against the average Russian communist, I don't remember anyone calling him a sissy.

For a final speculative look into the future I would

like to link together one of the earliest and one of the latest items in this history—Darwin, and the pill.

People have talked a good deal about the possible effects of the pill on society, and sexual relations, and the birth rate, and so on. There has been surprisingly little discussion about its possible genetic effects, and what there has been has been conducted mostly in 1984 terms, about the possibility of the state stepping in and stipulating which men and women should be allowed to breed and what type of citizen it wants to produce.

It is very unlikely to happen that way. Reproduction will continue to take place, as it has taken place since the days of the dinosaur and earlier, as a result of processes of natural selection. Only the pill will have thrown two monumental spanners into the works. One thing it will mean is that the evolutionary effects of natural selection may in some directions be immeasurably speeded up. The other thing is that slightly different types of human beings will be 'selected' as parents of the next generation.

Suppose that there is some genetic predisposition in certain women to be more favourably disposed than others to undertake the task of child-rearing. Such a predisposition has been treated in a previous chapter as a class and therefore a cultural difference, which to a great extent it probably is. But almost certainly there are also genetic factors involved. For instance, certain strains of poultry are more 'maternally' inclined than others, and this tendency can be greatly increased by selective breeding. A farmer who has invested in an incubator, and doesn't want his hens to stop laying eggs in order to sit on a clutch of them, can breed out the 'broodies' until he has eliminated this behaviour pattern entirely. He could also do the opposite, if it were in his economic interest to do so.

Back in the jungle or the sea or the savannah, a woman who was deficient in maternal promptings would be less likely than the average to perpetuate her line. She would continue to produce infants but would have less interest in them, less patience with them, and tend to neglect them. More of them would die, and the ones who survived would be unlikely to become dominant and prolific, though some might be taken under the wing of other females and thrive. The mechanism would be weighted appreciably against this non-maternal factor. It would not entirely die out, but its incidence would not increase.

In civilized society up to the last century the picture was different. Women who didn't want or like children continued to produce them quite prolifically because they fell in love, or because they wanted a home and security and marital status, and the children arrived as part of the package deal. The danger that they would actually die of neglect and starvation as a direct result of maternal indifference was less, and in the more prosperous sections of society where one woman produced the child and another woman reared it, it was nil. It was perfectly possible for a woman totally deficient in maternal promptings to produce a large, highly prosperous, and dominating line of progeny. Her kind would multiply, especially in the upper classes, and there is some reason to believe that it did: as with the poultry, the 'broodies' were increasingly bred out.

But if we arrive at a situation where a woman can have sex and security without having children, where children are a handicap to her in pursuing objectives more important to her, where nannies are a rare and terribly expensive luxury, and where demographers are plugging childlessness as a benefaction to humanity, such a woman is increasingly likely to have very few children or none. She will select herself out. It will not be a painfully slow and gradual business, as evolutionary processes have hitherto been, powered only by the fact that certain genetic factors make their inheritors marginally more or marginally less likely to survive. It could come down like a guillotine. If we lost the tradition that there is some 'status' involved in being a mother—it is a tradition beginning to falter and has recently for the first time ever been coming under direct fire—then the only women to have children would be the ones who cordially wanted them. The others would wipe themselves out in a generation.

It may well be that one hundred and fifty years hence people will read with astonishment of our fears that the net effect of the pill would be to defeminize women. Their own females will all be descendants of grandmothers and great-grandmothers so fizzing with oestrogen that a baby meant more to them than almost any other objective in life.

Any selective effect on the males would be far less instantaneous. The impetuous sexy Don Juan character who once careered around stamping his image over large areas of the countryside cannot do so from now on. He may still career, and his animal magnetism may prove as irresistible, but his likeness will not appear in the cradles for very much longer. Whether his type will die out depends on whether there is a hereditary element in his behaviour, or whether it is purely a psychological aberration, and we cannot be quite sure about this.

In the past husbands have been selected for a variety of reasons. Physical attractiveness is one, and fairly adaptive since it presupposes at least a degree of health and fitness. Being a 'good provider' is another, also adaptive since it implies at least a degree of competence. In the aristocracy 'breeding' has weighed heavily—genetically the worst bet of the lot since a noble name correlates neither with physical nor with mental viability. But the net genetic effect of all this in civilized society has been minimal, since unlike the gorilla and the baboon we have monogamy, and the prerogative of the 'breeding male' is unknown. Fatherhood is not limited to the handsome, the intelligent, the noble, or the dominant, as long as nearly everybody in the end gets married and children 'appear' as a consequence.

In future this may be slightly less true. The truly 'proletarian' family (literally those for whom their children constitute their only wealth and for the female her only status) is on the way out, and the woman with the 'lady's' attitude (that there are many other and easier ways of getting rewards out of life) is becoming the norm. In places where equality between the sexes has gone furthest, as for instance Moscow, the birth rate is going down fairly rapidly, not because of ecological exhortation by the state—the authorities are getting no joy out of the trend—but because more women have more options to choose from, and they make their own decisions on the matter.

Another tendency beginning to show itself in Russia and Scandinavia and other places is for girls of independent outlook to decide to have the baby without the husband. They obviously feel that the latter is a more bothersome thing to get saddled with than the former.

If both of these trends continue, then the process of husband-selecting might for the first time begin to have some genetic significance. The woman who decides to have a baby without a husband is making a cool and conscious choice anyway, and presumably doesn't select its father without thinking: 'I should be well content if my child turned out to resemble him.' And if, say, 15 per cent of women decided in this way against marriage, the remaining 85 per cent would have a wider choice and could afford to be more discriminating. Children are less likely to be the result of a woman's being 'swept off her feet' by an excess of passion. She can afford to get swept off her feet with joyous abandon for a year or so and still wait, before cementing the bond with a couple of children, to see whether the partnership looks like settling down comfortably for a long run; and the qualifications for this are somewhat different. It calls for less of sexiness on the male's part and more of loving-kindness. Men who possess most of this quality will be the likeliest to perpetuate their kind and help to form their children's minds.

What it adds up to is that, with the advent of the pill, woman is beginning to get her finger on the genetic trigger. What she will do with it we cannot quite foresee. But it is a far cry from the bull who gets to be prolific just because he's tops at beating the daylights out of all the other bulls.

It may be that for *homo sapiens* in the future, extreme manifestations of the behaviour patterns of dominance and aggression will be evolutionarily at a discount; and if that happens he will begin to shed them as once, long ago, he shed his coat of fur.

He may feel a little odd for the first few millennia because he is less accustomed to living without them than we are; but he has passed through more violent vicissitudes than this and survived. He is the most miraculous of all the creatures God ever made or the earth ever spawned. All we need to do is hold out loving arms to him and say:

'Come on in. The water's lovely.'

Postscript, 1985

That is the book I wrote fourteen years ago, in a mood of intense euphoria. Since then there have been many advances, both in the realms of women's studies and of evolutionary discoveries, but I have resisted the temptation to rewrite. I suspect that that would have destroyed some of the zing without adding much to compensate. Just one chapter has been omitted: it was based on Konrad Lorenz's contention that all species except *Homo sapiens* have built-in brakes on aggression against their own species. I no longer believe that to be true since reading Jane Goodall's account of chimpanzees stealing and eating one another's babies.

The book was an international best seller in its day and I received thousands of letters, the majority at first from women and many of them deeply moving. The women's movement was then just achieving lift-off, and I shall always be proud to have played some part in that continuing revolution. But over the years, as the trickle of women's books swelled into a flood and *The Descent of Woman* fell out of print, the letters became fewer and the balance changed. Recently most of them in effect have been asking: 'But whatever happened about the aquatic theory?' This postscript is intended as an answer to that question.

The book had three results that were both immediate and lasting. The first was that Hardy's theory was given a new lease of life. After sixteen years of silence most people, if they had heard of it at all, had forgotten it. Hardy himself had embarked on other projects that took up all his time, and no one else seemed inclined to carry a banner for it. It needed someone who either had unlimited supplies of

264 THE DESCENT OF WOMAN

moral courage or—as in my case—was an outsider with nothing to lose.

A second result was that up to that date it had been usual in books about prehistoric man to make reference to 'a hunting economy'. Professional anthropologists knew this to be a misnomer, but the knowledge was confined to a small minority: the popularizers used the term as though it were literally true. Judging by the letters, this issue above all aroused in women the indignant conviction that they had been systematically conned. Since that date, 'hunting/gathering economy' has become standard usage, and I suspect that the book made some contribution to the change.

The third change was silent but profound, and its effects are still being felt. In 1972 no one discussed the question of why *Homo sapiens* lost his body hair. The idea that he shed it in order to become cooler when chasing game had become virtually a platitude. Nobody challenged it or saw anything odd about it: it was a thing that every schoolboy knew. After 1972 the matter needed more delicate handling. New textbooks on evolution hesitated to repeat the platitude because they had lost confidence in it, but they were certainly not prepared to replace it with an aquatic explanation. So for several years most standard accounts of the evolution of man—including a couple of highly prestigious television series—side-stepped the whole issue. They discussed other ape-into-man changes—use of tools, bigger brain, upright posture, etc.—but never a word was breathed about hairlessness or the reasons for it. It was as though it had become, overnight, an unmentionable topic and, as we shall see, the search for the correct strategic response to this dilemma has passed through several subsequent phases, and still continues.

The initial critical response to *The Descent of Woman* was cordial: it was labelled as entertaining and thought provoking. No reviewer voiced an opinion on the plausibility or otherwise of the aquatic angle. Most of them felt that *that* was probably hokum, but they hesitated to commit themselves one way or the other until the experts had pronounced their verdict.

One reviewer for a periodical magazine held back his copy until the very last moment, waiting for the authoritative voices to speak up and crush these flimsy speculations under a weight of scientific evidence. When nobody spoke up, he ended his column by opining that this was because they were all 'too polite'.

Ten years later, with all the pro-aquatic arguments still unanswered, I was offered the same explanation in exactly the same words by an earnest young undergraduate: 'You see, they are all too polite.' Perhaps it is true. Locked away somewhere in the groves of academe there may be a complete refutation of the aquatic theory, the keys guarded night and day by a rota of scientific Galahads. According to the rules of their Order, upholding the truth is less important than the chivalrous principle which forbids them ever to contradict a lady . . . It is a charming idea, but I don't believe a word of it. I think it's a cop-out.

The first hurdle the book had to clear was the fact that I graduated in the wrong subject (English literature) and wrote plays for television. Now, there is nothing about the craft of fiction which automatically renders the writer incapable of accurate observation or logical thought. For example, one young writer born in 1866 was deeply interested in lichens and, after studying them closely for some years, she worked out that they must be a symbiotic association of fungi and algae. She was dead right, but when she reported her discovery at Kew she found she had been beaten to it by a short head. It was probably a stroke of luck for the botanists of her day that she wasn't a little quicker off the mark. Imagine those solid worthies being asked to accept a new scientific insight on the authority of the creator of Jemima Puddleduck and Mrs. Tiggywinkle! The mind boggles . . .

Even so, any resistance to the lichen discovery would have been short-lived, because it was merely adding one more brick to the edifice of accepted scientific knowledge, and no one feels threatened by that. The innovators who really raise hackles are the ones who tell them they have been on the wrong track all their lives—the ones who say: 'The truth is so simple and obvious that when you hear it you'll be kicking yourselves for not having seen it sooner.' (In practice they are much more likely to kick the innovator.)

The stock example of this type of innovator is Darwin, and the story is often told as if the resultant divisions of opinion were between the openminded scientists on the one hand and bigoted Churchmen on the other. That was by no means the case. The earliest and fiercest of Darwin's denouncers were themselves scientists—men like Adam Sedgewick and Richard Owen. Owen, head of the Natural History Department of the British Museum, had enjoyed the reputation of being Britain's leading biologist. His reaction to Darwin's book was vitriolic, and it was he in fact who urged on Bishop Wilberforce in the celebrated Oxford debate of June, 1860, and supplied him with the scientific information which went off, in the event, like a damp squib.

A more recent example shows that twentieth century scientists are still not immune to this human frailty. In 1912 the most important single item in the history of geology—the idea of continental drift—was advanced by a German called Alfred Wegener. In the 1920s his book on the subject had been published in five languages and was selling well, especially in Britain. Many laymen felt on reading it that a strong *prima facie* case had been made out, and waited to see what would happen.

They had a long wait. The greatest geological society in the world at that time, the Geological Society of London, ignored the idea—voiced no opinion, printed no comment, organized no debate. It was a gut reaction: they didn't want to know. Who was this Wegener, anyway? He wasn't even a geologist! He was a meteorologist, and they felt that the cobbler should stick to his last. (By the same token, the monk Gregor Mendel should have stuck to his prayers instead of laying the foundation of modern genetics, and A. R. Wallace should have carried on mounting his specimens instead of dead-heating with Darwin in his theory of Natural Selection, and zoologist Alister Hardy should not have encroached on the exclusive preserves of anthropologists.)

Much later, the Geological Society defended their policy of masterly inactivity by claiming that Wegener had already been answered, since a distinguished Cambridge mathematical physicist, Harold Jeffreys, in a brief reference in an appendix to a book entitled *The Earth*, had asserted that Wegener's theory was untenable. That should have settled the matter.

But contemporary reports make it clear that the rejection was not made on rational grounds. Physicist W. L. Bragg organized a meeting in Manchester to discuss continental drift and was staggered by the reception he got. 'The local geologists were furious; words cannot describe their utter scorn of anything so ridiculous as this theory.' Twenty-five years after the theory was first proposed, that attitude remained unchanged, and continental drift was written off as an exploded fallacy. After another twenty-five years it became the cornerstone of all contemporary thinking about the history of the earth.

All of which proves nothing about the aquatic theory, nor does it claim to. But it is a reminder that an establishment, however august, cannot put paid to an idea by spluttering at it. Nor can it be snowed under by politely declining to discuss it. It can only be killed by people who have considered it, have found reasons for rejecting it, and are prepared to state them. So the next step is to list the reasons that have been advanced to date of rejecting the Hardy thesis.

Some of the immediate knee-jerk reactions to the book in 1972 were understandable: they hadn't seen it coming. One anthropologist, who had published his own book on human evolution, was rung up by a journalist and asked to comment. He didn't need to read Morgan's book, he explained—he was familiar with Hardy's ideas and they were old stuff, and it was a matter of record that they had all been refuted years ago. Perhaps he believed what he was saying; perhaps he thought he had read something of the kind somewhere. His statement has since been echoed many times, always with total conviction, but no one when challenged has ever been able to remember where the alleged refutation was published, or who wrote it, or to quote any facts or arguments it is supposed to have contained.

Robert Ardrey, when questioned about *The Descent of Woman* in a newspaper interview, was loftily dismissive: no one, of course, would ever take it seriously. Elaine Morgan, he added, has obviously read my own books with some attention, but apart from that she doesn't appear to have read anything. Desmond Morris held his peace. Some years later, in his book *Man Watching*, he devoted five pages to Hardy's aquatic theory and openly praised it as a 'brilliant speculation'. He was one of the earliest writers to come out of the closet in this way and declare himself, if not quite convinced, at least open to conviction.

Live television produced some curious debating points. In a discussion on American TV one professor, confronted with Hardy's point about the orientation of body hair on an unborn baby, hazarded that to produce this effect the foetus must have been 'swimming round and round in the amniotic fluid'. He didn't explain how the dizzy creature escaped strangling itself in its own umbilical cord nor—since all unborn mammals are suspended in amniotic fluid —why only our own offspring should indulge in these antenatal gyrations. When asked by the interviewer about loss of body hair, he felt sure that a convincing explanation of this had once been given, by somebody or other, but for the moment both the explanation and the name of the explainer escaped him. I was afterwards assured that this professional could easily have demolished the aquatic theory if he hadn't been pulling his punches because I was a woman.

Again, I repeatedly encountered the argument about shell middens. 'If prehistoric man had been a diver, then we should have found shell middens.' Their absence was said to be of crucial significance. However, when it later transpired that prehistoric shell middens *had* been discovered (at Terra Amata) their significance abruptly shrank to nothing. It only proved that some specimens of *Homo erectus* had been able to dive—and what was surprising about that? Who had ever attempted to deny it?

As time passed, some of the data which made up the scientific consensus in the 1970s came to be modified. For example, there was the matter of the torrid Pliocene drought so luridly described by Robert Ardrey. We are now told that that picture was overdrawn: the climate was dry enough to shrink the forests and create the savannah, but would not have been harsh enough to force a water-dreading ape to take refuge in the sea. 'Does not this invalidate the whole theory?' I was asked.

No, because parallel with this new knowledge there have appeared further studies of Japanese macaques, teaching themselves to wade out into the sea with no fear of predators or drought to drive them. Their motive was only to clean the dirt from the food that was left for them on the shore by observers anxious to study their social interactions. The macaques soon discovered (like racoons) that dabbling in water was the most efficient cleansing method. They have been recorded on film, wading in up to their waists, standing upright to wash the food, and carrying it back up the beach again, walking bipedally just as Hardy had envisaged. Professor Lyall Watson has located estuaries on the east coast of Africa where he believes apes from the interior, millions of years ago, may well have begun to adapt to water in the same way.

From Leon P. La Lumiere of the Naval Research Department in Washington, DC, came another suggested location, in some ways even more persuasive. Researching the geological history of the north-eastern section of Africa, he made a connection between the aquatic theory and known episodes of flooding in that region. Large areas of that part of the continent were inundated and remained below sea level for very long periods. La Lumiere suggests that one place in particular, an area of high ground now known as the Danakil Horst—but at one time it would have been an island—provides a site where one population of apes might well have found themselves cut off, and so been compelled to adapt to a new life style.

This idea is attractive on three counts. To begin with, we now know that our line of descent splits off from that of the apes much more recently than was formerly believed, so that all the various differences between ourselves and them must have developed in an unusually short time. It is an evolutionary axiom—nowhere better shown than in Darwin's study of the finches on the Galapagos Islands—that islands are the places where evolution proceeds faster than anywhere else.

Furthermore, the picture conjured up on Danakil Island, from what we know about the climate and vegetation at that time, suggests that it may well have been partially fringed with salt marshes and semi-submerged forests—the precise type of habitat where today in Borneo we find the only other extant sea-going primate, the proboscis monkey.

The final reason why La Lumiere's theory is attractive is that it supplies the answer to a previously baffling question: if these primates adapted so well to life near and in the sea, why did they ever leave it? The geological record indicates that perhaps it was the other way round: the sea they swam in deserted *them*—became landlocked and finally dried up in the African sun, leaving the massive underground salt deposits that are there to this day.

Another assertion that became outdated concerned the length of the fossil gap. When I wrote, there was what anthropologists called the 'ten million year gap' in the fossil record between the last known fossils of ape-like *Ramapithecus* and the earliest ones of the manlike *Australopithecus*. As more discoveries were made, the fossil gap narrowed and now stands at something nearer to five million, or even less. The fact was triumphantly pointed out to me and I was asked whether I *still* had the gall to suggest that within that briefer space of time our ancestors had become semi-acclimatized to aquatic life and then re-adapted to life on land.

I replied that I am very happy about the five million years. Nothing has shortened the long dramatic list of differences between ourselves and our nearest anthropoid kin. The shorter the time in which these differences evolved, the more impossible it becomes to believe that, for the whole of that time, we were sharing the identical habitat of today's savannah chimpanzees, and that nothing distinguished our lives from theirs except a partial change of diet. The shorter the time, the more drastic the environmental disruption it would have required to modify us from a hairy ape into a creature in so many ways unique.

Very naturally, among scientists, the first to give serious attention to the theory were the young. Ever since the book appeared there have been undergraduates with the temerity to put up their hands and enquire whether there might not be something in it. Nowadays, on some campuses, they get the answer 'Maybe'. But the initial reaction was that this nuisance must now cease. One anthropologist complained in print that in every seminar there seemed to be this one awkward student 'in the back row'.

It is not difficult to gag young mavericks, especially if

you are popular and admired and carry enough guns. At that age, nobody likes being thought a fool. So one or two of the anthropological heavyweights slipped an *ex cathedra* judgement into whatever they were writing at the time, so that students should be left in no doubt of the correct attitude to adopt. Three or four lines were par for the course. For example, Stephen Jay Gould summed up: 'Elaine Morgan's *The Descent of Woman* is a speculative reconstruction of human prehistory from the women's point of view—and as farcical as more famous tall tales by and for men.'

Well, that was at least even-handed—as might have been expected, for Gould has never been anti-feminist. And it sounded final. At that point I ought to have accepted that the whole debate was right outside my league, and I should have thrown in the towel. Only two things prevented me: one minor question and one major one.

The minor question I asked myself was: Precisely which 'famous tall tales' for men is he talking about? Presumably the ones I set out to criticize—so presumably he shares my opinion of them. Why, then, did nobody of that calibre take the trouble to attach a similar warning to *those* books when they appeared, saying: 'Don't you believe a word of this. It's farcical'? A possible answer I arrived at was: 'It's not the sexism they are worried about—whether pro-male or pro-female. It's the Hardyism.' They were content with 'overheating in the hunt', even though they knew it was nonsense, because it stifled speculation and didn't endanger the rest of the story they had been telling all their lives. They got restive about the aquatic version because that would throw everything into the melting pot.

The major question was this: If the Tarzanist explanation of hairlessness and body fat and bipedalism, etc., is farcical, and if the Hardyist version is equally farcical, then why don't they simply give us the non-farcical explanation and settle the matter once and for all?

The answer is that they can't. They haven't got one.

Fair enough: the world is full of unanswered questions, and they are the life blood of science. But then they should have been presenting their students with this exhilarating challenge. They should have been saying: 'There are a lot of things we don't understand. Perhaps your generation will find the answers where we have failed.' Instead, they talked as if everything was clear to them, and that was misleading. It may have been unconscious rather than deliberate, but it was misleading.

I began to realize that by putting the aquatic theory into the same book as the feminism, and by the style in which it was written ('chipper', according to the *New Yorker*) I had made it too easy for the pundits to dismiss it out of hand without needing to give their reasons. There were broad hints that by handling it in this manner I had done more harm than good to Hardy's theory. I knew this was nonsense. Correspondence was coming in from all over the world to prove that, whereas it had been a dead letter, it was now alive and kicking; specialists in several different fields (paediatricians, dermatologists, marine biologists, etc.) were thinking about it and wondering whether it might have relevence to unsolved problems in their own particular fields.

I decided to restate the aquatic case separately, in a version which would not be quite as easy to brush aside. So I wrote another book called *The Aquatic Ape* (Souvenir Press, 1982) with a foreword by Sir Alister Hardy. In this the arguments were neutered and updated, and the style resolutely non-chipper. By this time new African fossils and fossil footprints had been discovered of pre-human creatures walking on two legs but leaving no traces of tools: so the contention that bipedalism was a by-product of carrying weapons had been quietly dropped by the Establishment. More was known about swimming babies, and diving, and dolphins; and the most ancient of the new fossil discoveries hinted at a location in the north-east of Africa for the original emergence of man. Protein dating had been introduced; the results reinforced the belief that all the differences between ourselves and the chimpanzees had evolved not gradually over 25 million years in jungle and/or grasslands, but rapidly, after a late split-off, during the five million year fossil gap, for reasons unknown.

Above all, a move had by then been made to formulate a third hypothesis to explain the ways in which human beings are unique. It featured the phenomenon of neoteny—the retention of infant characteristics into adult life. An unborn chimpanzee is hairless; a newborn chimpanzee's face is flatter and more humanoid than an adult's; and there were other more marginal resemblances, suggesting the possibility that man might be neither hunting ape nor swimming ape, but a sort of anthropoid Peter Pan—the ape that never grew up. However, this theory threw light only on the mechanism by which some of the major changes may have taken place: it had little or nothing to say about why they were necessary, or adaptive, or confined to one species.

The Aquatic Ape gave detailed comparisons of the three approaches (Savannah, Neoteny, Aquatic) and reprinted Hardy's original papers, and La Lumiere's Danakil theory. The reception it was given can best be described as circumspect. It has not been spluttered at—at least, not publicly—and a television film has now been made dealing with some of the points raised.

Meanwhile the campaign to discourage students from asking why they are not covered with fur continues unabated. Neoteny obviously did not quite do the trick, because a few years ago another stock professional response was dreamed up—that man is *not* hairless. It was pointed out that a human being has the same number of hair follicles over the body surface as a chimpanzee, and the same number of hairs, too, though diminished in size. If the actual numbers had altered, the argument goes, an explanation might well have been called for. But a mere modification in relative length need not necessarily be regarded as significant in evolutionary terms.

I wonder why Darwin puzzled so long about the giraffe. If he had only thought of counting, he would have found out that the giraffe had exactly the same number of necks as the zebra, so there was nothing that required explanation.

I have now heard this zany 'follicles' argument from the public platform, and on television, and from docile young anthropology students brainwashed by their tutors into thinking it settles the matter. It must surely be disingenuous. Nobody would claim that because a woman's face has the same number of follicles as a man's, there is no such thing as a beard. They are only able to get away with it because we are so accustomed to the sight of our own kind. Suppose a zoologist were confronted with a species of hairless bears, naked as Hollywood starlets, and asked to comment. He would not reply on the lines of: 'Very interesting. The muzzle is shorter than that of the nearest related species and the teeth have thicker enamel. I can see no other difference of any significance.' If he did say that, we should know he was either an idiot or was pulling our legs.

Naturally, not everyone is satisfied with the follicle argument. So, when the orthodox are forced to retreat yet further, they fall back on their last-ditch tactic. This consists of a vigorous counterattack, pouring scorn on the notion that all evolutionary changes can be explained in terms of adaptation and stressing that some of them are purely fortuitous, the result of random mutations. (Unfortunately there is no hard rule for distinguishing random changes from those resulting from natural selection: the expert expects us to accept his personal judgement on this question.)

The reference to chance is unconvincing. Mutations constantly occur in all species. If they are neutral in terms of survival-value (neither adaptive nor maladaptive) then indeed they may persist and survive, signifying nothing.

But these cases are marginal. They normally relate to the most trivial deviations, and are as rare as a tossed coin landing on its edge. Almost all random mutations come down heads or tails: if they are maladaptive in the habitat in which they arise, natural selection eliminates them; if they are adaptive, natural selection fosters them.

Hairlessness is very far from being a trivial feature, and in survival value it is never neutral. It is adaptive in water and maladaptive on land. However many times it might spontaneously arise in land-dwelling animals, whether by sheer chance or as part of a neotenic package, it would every time very quickly be bred out. Domestic cat breeders have recently taken advantage of just such a random mutation to establish a strain of hairless cats. In the wild such mutants must often have been born, but they would have been lucky to survive to puberty, and certainly none of these animals has ever engendered a successful species.

Moreover, hairlessness is only one among the remarkable cluster of human characteristics for which anthropologists have no agreed explanation. If all these arose by chance, we would surely be the most fortuity-prone creatures that ever walked the earth. Elsewhere in the plant and animal kingdoms the principle of neo-Darwinian natural selection continues to be freely and successfully applied: are they to be jettisoned in the case of this one species, in favour of chance and coincidence and mysteries that may never be solved? Why should reputable scientists be resorting to such convoluted arguments and still refusing even to give serious consideration to the Hardy alternative?

There is no mystery about that. Hardy himself first conceived the aquatic idea nearly sixty years ago. Like Darwin, he waited for decades before publishing anything. 'I wanted,' he says, 'to get a good professorship. I wanted to become a Fellow of the Royal Society, and I couldn't do that holding the aquatic theory. And so, quite candidly, I kept it smothered up.' His friends entreated him: 'Alister,

Alister, think of your reputation!' And when he finally spoke out he recalls thinking after the initial reception, 'Lord, I daren't go back to Oxford now.'

Desmond Morris is in no doubt of the pressures that operate. 'The orthodox anthropologist who has built up a theory of the evolution of man, omitting an aquatic phase, feels a bit foolish to even allow that it might have happened. He doesn't want this extra theory coming along and interfering with his nice story that he has been telling in the classroom for years. It is very, very difficult to admit that you have missed something as big as this . . . The marvellous thing about Alister Hardy is that he is a scientist with imagination. He is not a coward, as I'm afraid some academic scientists have become. You see, it's a very fine line between academic caution—proper academic caution—and craven academic cowardice.'

There is, however, a more positive side to this caution. The same anxiety that makes a career scientist wary of being the first to acclaim a startling new theory also makes him loth to be the last. Being praised today for 'soundness' sometimes ends in being condemned by posterity for bigotry. Probably all that is needed is a little more patience, and I have a strong feeling that the tide is already on the turn.

Bibliography

Ardrey, Robert, *African Genesis*. Collins, 1961.
—— *The Social Contract: A Personal Inquiry into the Evolutionary Source of Order and Disorder*. Collins, 1970.
—— *The Territorial Imperative: A Personal Inquiry into the Animal Origins of Property and Nations*. Collins, 1966.

Argyle, Michael, *The Psychology of Interpersonal Behaviour*. Penguin Books, 1970.

Barnett, Samuel A., *Instinct and Intelligence: Behaviour in Men and Animals*. MacGibbon Kee, 1967.

Bell, Peter R., ed., *Darwin's Biological Work*. Cambridge University Press, 1959.

Bertram, Colin, *In Search of Mermaids: The Manatees of Guiana*. London, Peter Davies, 1963.

Brecher, Ruth, and Brecher, Edward, eds., *An Analysis of Human Sexual Response*. Panther Books, 1969.

Calder, Nigel, *The Mind of Man*. B.B.C., 1970.

Cannon, Walter B., *Bodily Changes in Pain, Fear, Hunger and Rage*. 2nd edition. College Park, Md., McGrath, 1970.

Carrington, Richard, *Elephants: A short Account of Their Natural History, Evolution and Influence on Mankind*. London, Chatto & Windus, 1958.

Chance, Michael R., and Jolly, Clifford, *Social Groups of Monkeys, Apes and Men*. Cape, 1970.

Comfort, Alex, *Nature and Human Nature*. New York, Harper & Row, 1967.

Darwin, Charles, *The Expression of the Emotions in Man*

and Animals. Chicago, University of Chicago Press, 1965.

De Vore, Irven, ed., *Primate Behaviour: Field Studies of Monkeys and Apes*. New York, Holt, Rinehart & Winston, 1965.

—— and Eimerl, Sarel, *Primates*. New York, Time-Life Books, 1965.

Huxley, Julian, *Evolution: The Modern Synthesis*. Allen & Unwin, 1963.

Kinsey, Alfred C., et al., *Sexual Behaviour in the Human Female*. Philadelphia, W. B. Saunders, 1953.

Kohler, Wolfgang, *The Mentality of Apes*. 2nd revised edition. Penguin, 1959.

Lee, Richard, B., and De Vore, Irven, eds., *Man the Hunter*. Chicago, Aldine-Atherton, 1968.

Lorenz, Konrad, *On Aggression*. Methuen, 1966.

Masters, William H., and Johnson, Virginia E., *Human Sexual Response*. J. A. Churchill, 1966.

Milne, Louis J., and Milne, Margery, *The Senses of Animals and Men*. New York, Atheneum, 1962.

Montagu, M. F. Ashley, ed., *Culture and the Evolution of Man*. Oxford University Press, 1962.

Morris, Desmond, *The Human Zoo*. Cape, 1969.

—— *The Naked Ape: A Zoologist's Study of the Human Animal*. Cape, 1967.

Read, Leslie, *The Sociology of Nature*.

Roe, Anne, and Simpson, George G., eds., *Behaviour and Evolution*. Yale University Press, 1958.

Smith, Homer W., *From Fish to Philosopher*. Boston, Little, Brown & Co., 1953.

Tiger, Lionel, *Men in Groups*. New York, Nelson, 1970.

Williams, Leonard, *Man and Monkey*. Deutsch, 1967.

Index